Industrial Automation

About the Author
Frank Lamb has nearly 30 years of experience in the electrical and electronics industry and 20 years of experience in controls and automation. In 1996, he started Automation Consulting Services, Inc., an automation and systems integration company in Knoxville, Tennessee, concentrating on controls and panel building. In 2002, Automation Consulting Services began building smaller machines on-site and working with companies to install production lines. Mr. Lamb developed a multipurpose assembly and mistake-proofing machine known as the "SmartBench." His company now focuses on automation and Lean/Six-Sigma-oriented training and consulting.

Industrial Automation
Hands-On

Frank Lamb

New York Chicago San Francisco
Lisbon London Madrid Mexico City
Milan New Delhi San Juan
Seoul Singapore Sydney Toronto

Cataloging-in-Publication Data is on file with the Library of Congress.

McGraw-Hill Education books are available at special quantity discounts to use as premiums and sales promotions, or for use in corporate training programs. To contact a representative please visit the Contact Us page at www.mhprofessional.com.

Industrial Automation: Hands-On

Copyright ©2013 by McGraw-Hill Education. All rights reserved. Printed in the United States of America. Except as permitted under the United States Copyright Act of 1976, no part of this publication may be reproduced or distributed in any form or by any means, or stored in a data base or retrieval system, without the prior written permission of the publisher.

1 2 3 4 5 6 7 8 9 0   QFR/QFR   1 2 0 9 8 7 6 5 4 3

ISBN   978-0-07-181645-8
MHID   0-07-181645-3

The pages within this book were printed on acid-free paper.

Sponsoring Editor	**Project Manager**	**Production Supervisor**
Michael Penn	E. Grishma Fredric, Newgen Knowledge Works Pvt Ltd	Pamela A. Pelton
Acquisitions Coordinator		**Composition**
Amy Stonebraker	**Copy Editor**	Newgen Knowledge Works Pvt Ltd
	Robin O'Dell	
Editorial Supervisor		**Art Director, Cover**
David E. Fogarty	**Proofreader**	Jeff Weeks
	MEGAS	

Information contained in this work has been obtained by McGraw-Hill Education from sources believed to be reliable. However, neither McGraw-Hill Education nor its authors guarantee the accuracy or completeness of any information published herein, and neither McGraw-Hill Education nor its authors shall be responsible for any errors, omissions, or damages arising out of use of this information. This work is published with the understanding that McGraw-Hill Education and its authors are supplying information but are not attempting to render engineering or other professional services. If such services are required, the assistance of an appropriate professional should be sought.

Contents

Preface xiii

1 Automation and Manufacturing 1
- 1.1 Automation 1
  - 1.1.1 Advantages 2
  - 1.1.2 Disadvantages 2
  - 1.1.3 The Factory and Manufacturing 3
  - 1.1.4 The Manufacturing Environment 6

2 Important Concepts 9
- 2.1 Analog and Digital 9
  - 2.1.1 Scaling 10
- 2.2 Input and Output (Data) 11
  - 2.2.1 Discrete I/O 11
  - 2.2.2 Analog I/O 12
  - 2.2.3 Communications 15
  - 2.2.4 Other Types of I/O 21
- 2.3 Numbering Systems 22
  - 2.3.1 Binary and BOOL 22
  - 2.3.2 Decimal 22
  - 2.3.3 Hexadecimal and Octal 22
  - 2.3.4 Floating Point and REAL 24
  - 2.3.5 Bytes and Words 24
  - 2.3.6 ASCII 24
- 2.4 Electrical Power 25
  - 2.4.1 Frequency 25
  - 2.4.2 Voltage, Current, and Resistance .. 25
  - 2.4.3 Power 27
  - 2.4.4 Phase and Voltages 27
  - 2.4.5 Inductance and Capacitance 28
  - 2.4.6 Solid-State Devices 29
  - 2.4.7 Integrated Circuits 31
- 2.5 Pneumatics and Hydraulics 32
  - 2.5.1 Pneumatics 33
  - 2.5.2 Hydraulics 34
  - 2.5.3 Pneumatic-Hydraulic Comparison ... 35

v

Contents

- 2.6 Continuous, Synchronous, and Asynchronous Processes ... 36
  - 2.6.1 Continuous Processes ... 36
  - 2.6.2 Asynchronous Processes ... 36
  - 2.6.3 Synchronous Processes ... 36
- 2.7 Documentation and File Formats ... 36
  - 2.7.1 Drafting and CAD ... 37
  - 2.7.2 Other Design Packages and Standards ... 39
  - 2.7.3 Image File Formats ... 42
- 2.8 Safety ... 44
  - 2.8.1 Hazard Analysis ... 47
  - 2.8.2 Emergency Stops ... 48
  - 2.8.3 Physical Guarding ... 51
  - 2.8.4 Lockout/Tagout ... 51
  - 2.8.5 Design Mitigation ... 53
  - 2.8.6 Guard Devices ... 53
  - 2.8.7 Software ... 55
  - 2.8.8 Intrinsic Safety ... 56
- 2.9 Overall Equipment Effectiveness ... 57
  - 2.9.1 Availability ... 58
  - 2.9.2 Performance ... 58
  - 2.9.3 Quality ... 59
  - 2.9.4 Calculating OEE ... 59
- 2.10 Electrostatic Discharge ... 60

3 Components and Hardware ... 61
- 3.1 Controllers ... 61
  - 3.1.1 Computers ... 61
  - 3.1.2 Distributed Control Systems (DCSs) ... 62
  - 3.1.3 Programmable Logic Controllers (PLCs) ... 62
  - 3.1.4 Embedded Controllers and Systems ... 64
- 3.2 Operator Interfaces ... 65
  - 3.2.1 Text-Based Interfaces ... 66
  - 3.2.2 Graphical Interfaces ... 66
  - 3.2.3 Touch screens ... 67
- 3.3 Sensors ... 69
  - 3.3.1 Discrete Devices ... 70
  - 3.3.2 Analog ... 77
  - 3.3.3 Special Purpose Sensors ... 84
- 3.4 Power Control, Distribution, and Discrete Controls ... 93

		3.4.1	Disconnects, Circuit Breakers, and Fusing	94
		3.4.2	Distribution and Terminal Blocks	98
		3.4.3	Transformers and Power Supplies	100
		3.4.4	Relays, Contactors, and Starters	102
		3.4.5	Timers and Counters	104
		3.4.6	Push Buttons, Pilot Lights, and Discrete Controls	107
		3.4.7	Cabling and Wiring	110
	3.5	Actuators and Movement		115
		3.5.1	Pneumatic and Hydraulic Actuators and Valves	115
		3.5.2	Electric Actuators	119
		3.5.3	Motion Control	119
	3.6	AC and DC Motors		121
		3.6.1	AC Motors	122
		3.6.2	DC Motors	126
		3.6.3	Linear Motors	129
		3.6.4	Servomotors and Stepper Motors	129
		3.6.5	Variable Frequency Drives	132
	3.7	Mechanisms and Machine Elements		134
		3.7.1	Cam-Driven Devices	135
		3.7.2	Ratchet and Pawl Systems	136
		3.7.3	Gearing and Gear Reduction	136
		3.7.4	Bearings and Pulleys	141
		3.7.5	Servomechanisms	143
		3.7.6	Ball Screws and Belt-Driven Linear Actuators	144
		3.7.7	Linkages and Couplings	145
		3.7.8	Clutches and Brakes	147
	3.8	Structure and Framing		148
		3.8.1	Steel Framing	148
		3.8.2	Aluminum Extrusion	153
		3.8.3	Piping and Other Structural Systems	153
		3.8.4	Electrical Enclosures and Ratings	154
4	**Machine Systems**			**165**
	4.1	Conveyors		165
		4.1.1	Belt Conveyors	166
		4.1.2	Roller Conveyors	167
		4.1.3	Chain and Mat Conveyors	169
		4.1.4	Vibrating Conveyors	170
		4.1.5	Pneumatic Conveyors	171
		4.1.6	Accessories	171

4.2	Indexers and Synchronous Machines		172
	4.2.1	Rotary Cam Indexers	172
	4.2.2	Synchronous Chassis Pallet Indexers	172
	4.2.3	Walking Beams	173
	4.2.4	Pick-and-Place	174
4.3	Part Feeders		175
	4.3.1	Vibratory Bowls and Feeders	175
	4.3.2	Step and Rotary Feeders	176
	4.3.3	Escapements and Parts Handling	177
4.4	Robots and Robotics		178
	4.4.1	Articulated Robots	178
	4.4.2	SCARA Robots	179
	4.4.3	Cartesian Robots	180
	4.4.4	Parallel Robots	180
	4.4.5	Robot Basics and Terminology	181
	4.4.6	Robot Coordinate Systems	184

5 Process Systems and Automated Machinery 185

5.1	Chemical Processing		186
5.2	Food and Beverage Processing		187
5.3	Packaging		188
5.4	Web Handling and Converting		190
5.5	Metal, Plastic, Ceramic, and Glass Processing		192
	5.5.1	Metals	192
	5.5.2	Plastics	198
	5.5.3	Ceramics and Glass	206
5.6	Assembly Machines		208
	5.6.1	Part Handling	208
	5.6.2	Fastening and Joining	210
	5.6.3	Other Assembly Operations	213
5.7	Inspection and Test Machines		214
	5.7.1	Gauging and Measurement	214
	5.7.2	Leak and Flow Testing	214
	5.7.3	Other Testing Methods	217

6 Software 219

6.1	Programming Software		219
	6.1.1	Programming Concepts	220
	6.1.2	Programming Methodologies	221
	6.1.3	Languages	225
6.2	Design Software		240
6.3	Analysis Software		240
6.4	Office Software		241
6.5	SCADA and Data Acquisition		242

	6.6	Databases and Database Programming	243
	6.7	Enterprise Software	245
7	**Occupations and Trades**		**251**
	7.1	Engineering	251
		7.1.1 Mechanical	252
		7.1.2 Electrical and Controls	252
		7.1.3 Industrial and Manufacturing Engineering	253
		7.1.4 Chemical and Chemical Process Engineers	254
		7.1.5 Other Engineering Disciplines and Job Titles	255
	7.2	Trades	257
		7.2.1 Mechanical	257
		7.2.2 Electrical	260
8	**Industrial and Factory Business Systems**		**263**
	8.1	Automation-Related Businesses	263
		8.1.1 Manufacturers	263
		8.1.2 OEMs	264
		8.1.3 Manufacturers' Representatives	264
		8.1.4 Distributors	264
		8.1.5 Machine Builders	264
		8.1.6 Systems Integrators	265
		8.1.7 Consultants	265
	8.2	Departments and Functions	265
		8.2.1 Management	266
		8.2.2 Sales and Marketing	267
		8.2.3 Engineering and Design	267
		8.2.4 Maintenance	268
		8.2.5 Manufacturing and Production	269
		8.2.6 Finance and Human Resources	270
		8.2.7 Quality	271
		8.2.8 Information Technology	273
	8.3	Lean Manufacturing	273
		8.3.1 Kanban and "Pull"	275
		8.3.2 Kaizen	276
		8.3.3 Poka-Yoke	277
		8.3.4 Tools and Terms	278
	8.4	Systemization	281
		8.4.1 Job and Task Descriptions	282
		8.4.2 Communications	283
		8.4.3 Hiring and Training	284
		8.4.4 Engineering and Project Notebooks	285

Contents

9 Machine and System Design **287**
- 9.1 Requirements 287
  - 9.1.1 Speed 287
  - 9.1.2 Improvements 288
  - 9.1.3 Cost 288
  - 9.1.4 Requirements Documentation 289
- 9.2 Quoting 289
  - 9.2.1 Quote Request 289
  - 9.2.2 Quote Analysis 290
  - 9.2.3 The Decision 291
- 9.3 Procurement 291
  - 9.3.1 Terms 291
- 9.4 Design 292
  - 9.4.1 Mechanical 292
  - 9.4.2 Electrical and Controls 294
  - 9.4.3 Software and Integration 296
- 9.5 Fabrication 298
  - 9.5.1 Structural 298
  - 9.5.2 Mechanical 298
  - 9.5.3 Electrical 298
  - 9.5.4 Assembly 299
- 9.6 Start-up and Debug 300
  - 9.6.1 Mechanical and Pneumatics 300
  - 9.6.2 Packaging Integration 300
  - 9.6.3 Controls 301
- 9.7 FAT and SAT 302
  - 9.7.1 Factory Acceptance 302
  - 9.7.2 Site Acceptance 302
- 9.8 Installation 303
  - 9.8.1 Shipping 303
  - 9.8.2 Contract Millwright and Electrician . 303
- 9.9 Support 304
  - 9.9.1 The First Three Months 304
  - 9.9.2 Warranty 304

10 Applications **305**
- 10.1 Binder-Processing Machine 305
- 10.2 Crystal Measurement 306
- 10.3 SmartBench 308
- 10.4 Sagger Load Station 310
- 10.5 Tray Handlers 312
- 10.6 Cotton Classing System 313

A	ASCII Table	317
B	Ampacity	326
C	Motor Sizing	327
D	NEMA Enclosure Tables	332
E	Manufacturers, Machine Builders, and Integrators	336
F	Thermocouples	341
Bibliography		345
Index		347

Preface

This book is meant to serve as an outline for automation and industrial machinery concepts and terminology. It is suitable as a guidebook for newcomers to the field of automation as well as a reference for the seasoned automation professional. The book emphasizes control systems, but many other subjects—including machine building, mechanical engineering and devices, manufacturing business systems, and job functions in an industrial environment—are also covered extensively.

I began my career in the U.S. Air Force as an electronics instructor and engineering/installation technician. The military requires its members to follow detailed instructions and document work activities and procedures accurately—a requirement that equipped me with the tools and discipline necessary to pursue a career in engineering. I learned the fundamental elements of electronics and electricity from a practical, hands-on perspective—tracing signal flow through schematics and using specialized tools such as soldering irons and wire-wrap guns to install and repair components. I also learned to develop, follow, and present lesson plans in a military classroom. During my service in the military, I was fortunate enough to be able to travel internationally and experience other cultures and methods of accomplishing things.

After eight years in the military, I started college as a 30-year-old freshman. My math and science background was not particularly strong, so I was forced to play catch-up to keep up with my student peers who came out of high schools and junior colleges with some calculus and physics experience. Fortunately my military background gave me the self-discipline to study these prerequisites independently, and I began to learn the core subjects of engineering. I attended a large state university with a highly regarded engineering department and numerous top-notch professors. Although all engineering students were required to complete mechanical and industrial engineering courses, my interest and concentration was in the controls field of electrical engineering. To supplement my controls classes,

I also studied power and digital electronics, plasma, communications, drafting/CAD, thermodynamics, semiconductor theory, and various computer programming classes in addition to an assortment of liberal arts and general courses. This curriculum provided a well-rounded and complete engineering education intended to ready me for entry into the workforce.

After finishing college and obtaining a degree in electrical and computer engineering, I discovered that there was a gap in my education from a practical standpoint. Industrial automation is tricky. The theoretical requirements are great, but the practical knowledge necessary is even greater. The traditional ways of teaching automation simply don't transfer the *practical* knowledge you obtain after years of experience. Sure, you can learn a lot of mathematical and scientific concepts that will give you an excellent *theoretical* understanding of the field, but they fail to provide the necessary practical knowledge that normally comes from years of trial and error. This book is intended to fill some of these "experience gaps."

My automation background has been primarily in machine building and systems integration. Before starting my own custom systems integration and machine-building company, I worked for two controls and electrical component distributers and learned the value of manufacturers' catalogs and training classes. Much of the information contained in this book is gleaned from catalogs that provide specifications of equipment for systems. Prior to the emergence of the Internet as a widespread and accessible resource, most of the technical data for components had to be obtained from specification sheets and physical catalogs. Classes and seminars presented by manufacturers were—and still are—excellent resources for hands-on training using actual hardware.

After operating a small machine-building and systems integration company for 10 years, I went to work for a large custom machine builder. This provided me with insight about how major corporations and engineering firms bid out and procure large integrated turnkey production lines and systems and how teams of engineers work together on extremely complex systems to produce integrated production lines. The many valuable tools and templates I used in my position at Wright Industries were instrumental in my ability to design systems in a coordinated and organized way. I will always be grateful to Wright Industries and the Doerfer Companies for the experiences and training I received during my tenure there, as well as the ability to play with large-scale and expensive "toys."

While working within this much larger organization, I was often asked how things worked or what the best technique might be to accomplish a particular task. I also had questions myself on areas outside my expertise. As these important questions were answered, I began collecting information with the primary aim of starting a general guidebook for these often-asked questions in the field of automation. I also started a blog, www.automationprimer.com, and began posting subjects of interest to the automation community. This allowed me to refine some of the subjects and begin organizing the content of this book. It also allowed me to cross-link with several other automation bloggers and gain valuable contacts in the industry.

During this period I also obtained my Six Sigma Green Belt certification. My interest in lean manufacturing and the business aspect of the automation and manufacturing industries grew, and I began adding more business-related content to the book. I left Wright Industries in 2011 and restarted my automation company with more of a focus on the education and consulting aspect of industrial automation rather than machine design and programming.

One of the major purposes of this book is to serve as a single-point resource for those involved in the design and use of automated machinery. I used many of the charts and tables in this book as design aids when specifying systems during my automation career, and I hope they serve as useful quick-reference guides for readers as well. There are also many topics that provide general information on industry-related subjects that might also be of interest to readers who hope to expand their general knowledge of automation-related subjects.

The book is laid out in an outline format for easy reference. Chapter 1 provides a general overview of manufacturing and automation. Chapter 2 introduces many of the concepts used in automation, controls, machinery design, and documentation. Chapter 3 discusses many of the individual hardware components used in the automation industry. Chapter 4 links some of these components together and describes some of the machinery subsystems that help comprise an automated machine or line. Chapter 5 brings these subsystems together and exemplifies machinery used in some of the different areas of manufacturing. Chapter 6 covers some of the different kinds of software used in the programming, design, and documentation of industrial machinery and information systems as well as business enterprise software. Chapter 7 describes job functions in the automation and manufacturing industries, and Chap. 8 covers some of the business organization and concepts used in the industrial and manufacturing fields. Lean manufacturing and various business tools are also discussed there. Chapter 9 covers a hypothetical machine procurement, design, and implementation process, while Chap. 10 contains some examples of automation projects and systems I have been involved in during my career. There are also a number of handy tables and charts and an index in the back of the book.

Mechanical engineers who want to know more about controls or business, electrical engineers and technicians who need more information on mechanical concepts and components, and factory management employees needing more background on technical subjects will find this book helpful. Machine operators hoping to move into the maintenance field and maintenance technicians needing more information on engineering techniques will also find subjects of interest contained within these pages.

Because the subject matter is broad, covering both technical and business aspects of manufacturing and industrial automation, no particular subject is covered in great depth. There are thousands of excellent resources available in textbooks, catalogs, and online that go into much more detail on specific areas of the automation, business, and manufacturing fields. I would encourage readers to explore these subjects in greater detail and hope that you get as much enjoyment out of this fascinating field as I have.

This material has been developed with the assistance of many individuals to whom I wish to express my sincere appreciation. To my daughter Mariko, who provided extensive editing, proofreading, and other book- and picture-related help throughout the entire process. To the technical reviewers who made suggestions and corrections in their areas of expertise:

> Tony Bauer—software developer, founder of factoryswblog.org
>
> John Bonnette, MBA—mechanical engineer, reliability manager at DSM Dyneema
>
> Jeff Buck, PE—electrical and mechanical engineer, vice president at Automation nth
>
> Trent Bullock—E and I technician at DSM Dyneema
>
> Jason Gill—industrial engineer, lean Six Sigma manager at Mayekawa USA
>
> Gordon Holmes—senior project engineer/controls at Wright Industries
>
> Michael Lee—welder at Mayekawa USA
>
> Ron Lindsey—software, robotics, and vision engineer at Wright Industries
>
> Bill Martin—president at Martin Business Consulting
>
> Tom Nalle—president at Nalle Automation Systems
>
> Charlie Thi Rose—mechanical engineer, president at C. T. Rose Enterprises
>
> Louis Wacker—senior mechanical design engineer at Wright Industries

Finally, I would like to acknowledge the encouragement and patience of my wife, Mieko, who helped to make this book possible.

Nashville, Tennessee   F. B. L.

CHAPTER 1
Automation and Manufacturing

Human beings have been making things for many thousands of years. Originally most products were made on an individual as-needed basis; if a tool was required, it was fashioned by hand and in turn used to make more tools. As time passed, more complex techniques were developed to help people accomplish fabrication and production tasks. Metalworking technology, weaving looms, water-driven grinding mills, and the development of steam and gasoline engines all contributed to a greater ability to make various products, but things were still generally made one at a time by craftspeople skilled in various techniques. It was only after the Industrial Revolution and common use of electrical energy and mechanisms that manufacturing of products on a large scale became commonplace.

1.1 Automation

Automation is the use of logical programming commands and mechanized equipment to replace the decision making and manual command-response activities of human beings. Historically, mechanization—such as the use of a timing mechanism to trip a lever or ratchet and pawl—aided humans in performing the physical requirements of a task. Automation, however, takes mechanization a step further, greatly reducing the need for human sensory and mental requirements while simultaneously optimizing productivity.

It is believed that the term *automation* was first coined in the 1940s by a Ford Motor Company engineer describing various systems where automatic actions and controls were substituted for human effort and intelligence. At this time, control devices were electromechanical in nature. Logic was performed by means of relays and timers interlocked with human feedback at decision points. By wiring relays, timers, push buttons, and mechanical position sensors together, simple logical motion sequences could be performed by turning on and off motors and actuators.

With the advent of computers and solid-state devices, these control systems became smaller, more flexible, and less expensive to implement and modify. The first programmable logic controllers were developed in the 1970s and 1980s by Modicon in response to a challenge by GM to develop a substitute for hardwired relay logic. As technology improved and more automation companies entered the market, new control products were developed. Today, myriad computerized logic control devices developed by hundreds of different manufacturers exist in the industry.

1.1.1 Advantages

A few advantages of automation are:

- Human operators performing tasks that involve hard physical or monotonous work can be replaced.
- Human operators performing tasks in dangerous environments, such as those with temperature extremes or radioactive and toxic atmospheres, can be replaced.
- Tasks that are beyond human capabilities are made easier. Handling heavy or large loads, manipulating tiny objects, or the requirement to make products very quickly or slowly are examples of this.
- Production is often faster and labor costs less on a per product basis than the equivalent manual operations.
- Automation systems can easily incorporate quality checks and verifications to reduce the number of out-of-tolerance parts being produced while allowing for statistical process control that will allow for a more consistent and uniform product.
- Automation can serve as the catalyst for improvement in the economies of enterprises or society. For example, the gross national income and standard of living in Germany and Japan improved drastically in the 20th century, due in large part to embracing automation for the production of weapons, automobiles, textiles, and other goods for export.
- Automation systems do not call in sick.

1.1.2 Disadvantages

Some disadvantages of automation are:

- Current technology is unable to automate all desired tasks. Some tasks cannot be easily automated, such as the production or assembly of products with inconsistent component sizes or in tasks where manual dexterity is required. There are some things that are best left to human assembly and manipulation.

- Certain tasks would cost more to automate than to perform manually. Automation is typically best suited to processes that are repeatable, consistent, and high volume.
- The research and development cost of automating a process is difficult to predict accurately beforehand. Since this cost can have a large impact on profitability, it is possible to finish automating a process only to discover that there is no economic advantage in doing so. With the advent and continued growth of different types of production lines, however, more accurate estimates based on previous projects can be made.
- Initial costs are relatively high. The automation of a new process or the construction of a new plant requires a huge initial investment compared with the unit cost of the product. Even machinery for which the development cost has already been recovered is expensive in terms of hardware and labor. The cost can be prohibitive for custom production lines where product handling and tooling must be developed.
- A skilled maintenance department is often required to service and maintain the automation system in proper working order. Failure to maintain the automation system will ultimately result in lost production and/or bad parts being produced.

Overall, the advantages would seem to outweigh the disadvantages. It can be safely said that countries that have embraced automation enjoy a higher standard of living than those that have not. At the same time, a concern is often aired that automating tasks takes jobs from people that used to build things by hand. Regardless of the social implications, there is no doubt that productivity increases with the proper application of automation techniques.

1.1.3 The Factory and Manufacturing

A *factory*, or manufacturing plant, is an industrial building where workers produce, assemble, process, or package goods by operating and supervising machines and processing lines (Fig. 1.1). Most modern factories house innovative machinery used for production, gauging, testing, packaging, and a host of other manufacturing-related operations. From a business perspective, factories serve as the central site where labor, capital, and plant are concentrated for the development of mass produced, small batch, or specialty goods.

The factory setting proved to be an efficient environment for mass production during the Industrial Revolution, when England spearheaded a shift from an agrarian-based society to one powered

Chapter One

FIGURE 1.1 Factory.

by machinery and manufacturing. At this time, factories simply served as buildings where laborers gathered to produce goods using simple tools and machinery. Advancements in agriculture and textile- and metal-manufacturing technologies coupled with cheap labor resulted in increased output, efficiency, and profit for factory owners.

In the early 20th century, Henry Ford advanced the factory concept further with the innovation of mass production in order to meet a growing demand for his Model T automobiles. Through a combination of the employment of precision manufacturing, the division of highly specialized labor, the use of standardized and interchangeable parts, and the creation of a continuously rolling, precisely timed assembly line, Ford was able to drastically reduce assembly time per vehicle and ultimately decrease production costs.

Ford's model changed the way virtually all goods were manufactured in the 20th century and paved the way for the next generation of factories to evolve with several improvements.

One such improvement, pioneered by American mathematician William Edwards Deming, was the advancement of statistical methods of quality control—an innovation he brought overseas, which turned Japanese factories into world leaders in cost-effectiveness and production quality. The latest advancements in quality control have led to the concepts of Six Sigma and lean manufacturing. These concepts are covered in depth in the later chapters of this book.

Another improvement to the factory model was the innovation of industrial robots, which began appearing on factory floors in the 1970s. These computer-controlled arms and grippers conducted simple tasks but were instrumental in improving speed and cutting costs. Primary functions of these high-endurance precision machines include welding, painting, pick and place, assembly, inspection, and testing.

Automation and Manufacturing 5

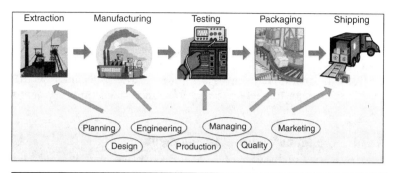

FIGURE 1.2 Production cycle.

Manufacturing is the systematic fabrication of goods through the use of machines, tools, and labor. Manufacturing of the 21st century most commonly applies to industrial production, where large quantities of raw materials are transformed into finished products. This process usually happens in many stages; a finished product from one process becomes one of many components required for another process. Those finished products may be sold to end users by way of wholesalers or retailers, or they can be used to manufacture other, more complex products before they are ultimately sold to consumers. Assembly, conversion, packaging, and processing/batching are all examples of manufacturing operations.

Figure 1.2 illustrates a simple diagram of production for a typical product. Note that for every stage of the production cycle, there are many support activities involved that do not directly affect the manufacturing process itself. Many of these activities are also performed by separate companies and facilities, involving financial transactions and product movement between corporations and locations.

Most industrial automation occurs within the manufacturing and production phase of the product life cycle; however, examples of automated functions can be found in many nonindustrial applications. As an example, virtually every computer program uses command and response automation to operate; the word *automation* also refers to the application of computer processing to a task, an example being the term *office automation*.

Additive manufacturing, also known as 3-D printing, is a process of making solid objects from a solid model drawing by incrementally adding successive layers of material rather than removing them by machining, which is a subtractive process. While not currently practical for mass production from a time and cost standpoint, it is a useful method for making a single object for rapid prototyping.

Methods used in additive manufacturing include extruding layers of polymer or metal; laminating layers of foil, paper, or plastic film; or using electron beams to selectively fuse granular metals layer by layer. Related methods of rapid prototyping are stereolithography and digital light processing. These methods produce solid parts out of a liquid by exposing the liquid to intense light beams, hardening the exposed polymer.

Additive manufacturing is a rapidly evolving technology that may have a major impact on the future of manufacturing production methods.

1.1.4 The Manufacturing Environment

Factories are often thought of as dirty, noisy places with lots of heavy equipment; this is not always true. Places like foundries and metal-processing plants can certainly have these characteristics, requiring ear and clothing protection; however, factories can also be immaculate and relatively quiet, depending on what is being produced or processed.

Most industrial facilities require a minimum of protective equipment to be worn while on the factory floor. Safety glasses are nearly always required. There is often a box containing extra pairs for visitors next to the doors from the office space to the plant. Earplug dispensers are also often located in the same area. In food-processing plants and clean rooms, hairnets and beard nets, coats, shoe covers, and gloves may also be required.

A clean room is an environment where particles of dust or contaminants are prevented from entering. Clean rooms are classified by the number of particles above certain sizes that may be allowed within a volume of air. The ISO standard on clean rooms was published in 1999, establishing numbers for classification as shown in Table 1.1.

This standard is used as the basis for the European Union classifications. In the United States, Federal Standard 209D is used to classify clean rooms in a similar way, as shown in Table 1.2. This classification defines the number of particles in a cubic foot of air rather than a cubic meter.

Federal Standard 209E uses the same criteria as 209D, but it also defines classes by cubic meter in addition to cubic foot. Since most U.S. facilities still refer to clean room classifications as "class 100,000" and below, Standard 209E is not shown here.

Controlled environments are also referred to by grade when referring to antiseptic or sterilized conditions. Grades A through D are used in conjunction with clean room classes with grade "A" being the most restrictive, used in aseptic preparation and filling of sterilized products.

Clean rooms are used in manufacture of semiconductor devices, computer hard drives, pharmaceuticals, and some food-preparation facilities. Additional methods of maintaining clean room quality include minimum air volume exchange rates, decontamination, air locks, and filtering. Components intended for use in a clean room

Classification Numbers (N)	Maximum Concentration Limits (particles/cubic meter of air) for Particles Equal to and Larger Than the Considered Sizes Shown Below (micrometers)					
	0.1	0.2	0.3	0.5	1	5
ISO 1	10	2				
ISO 2	100	24	10	4		
ISO 3	1000	237	102	35	8	
ISO 4	10,000	2370	1020	352	83	
ISO 5	100,000	23,700	10,200	3520	832	29
ISO 6	1,000,000	237,000	102,000	35,200	8320	293
ISO 7				352,000	83,200	2930
ISO 8				3,520,000	832,000	29,300
ISO 9				35,200,000	8,320,000	293,000

TABLE 1.1 BS EN ISO Standard, 14644–1 "Classification of Air Cleanliness"

	Measured Particle Size (Micrometers)				
Class	0.1	0.2	0.3	0.5	5.0
1	35	7.5	3	1	NA
10	350	75	30	10	NA
100	NA	750	300	100	NA
1,000	NA	NA	NA	1000	7
10,000	NA	NA	NA	10,000	70
100,000	NA	NA	NA	100,000	700

TABLE 1.2 Federal Standard 209D Class Limits

environment typically require extra preparation before being approved for use. Using low outgassing materials, covering all moving parts that might generate debris, using greases that do not splatter, or even pulling a slight vacuum within the component so that contaminants are exhausted out of the room are all methods of component preparation. Robots and ball screw stages are examples of clean room ready components.

An additional concern in manufacturing plants is the use and disposal of hazardous substances. For chemicals used in manufacturing facilities, a Material Safety Data Sheet (MSDS) or Safety Data Sheet (SDS) is required for cataloguing information on substances. Information on how to handle or work with the substance; physical data such as melting, boiling, or flash points; storage; disposal; and spill-handling procedures are all required. Formats can differ, depending on national and local regulations. These data sheets are placed close to the location where any chemical is stored or used.

In the electronics industry, dangerous substances are prohibited above certain concentrations. Materials such as lead, mercury, cadmium, and several other substances are limited by weight or percentage, and products are tested for compliance. The Restriction of Hazardous Substances (RoHS) directive limits amounts of dangerous chemicals in consumer products and packaging. Products affected include electronic and electrical tools, automatic dispensers, lightbulbs and lighting equipment, household appliances, toys, and many other consumer devices.

Discharge of waste substances into water, landfills, and air is also regulated by various U.S. and international agencies. While often criticized by environmental groups, most companies in industrialized nations spend vast sums of money ensuring that factory impacts on the environment are minimized or eliminated. Water treatment, waste packaging, and air filters and scrubbers are used to reduce the impact of pollutants on the environment. Remediation of polluted areas is also an important requirement of manufacturing and industry enforced by governmental agencies.

CHAPTER 2
Important Concepts

2.1 Analog and Digital

The most basic element of automation logic is its *digital* state. A switch or signal may only be on or off. This can be represented as a signal being a 0 (off) or a 1 (on). There are many elements in an automation scheme that can be represented as a 1 or 0—the state of a switch or sensor; the state of a motor, valve, or pilot light; or even the state of a machine itself.

The state of many devices cannot be so simply described. A motor can be described as being on or off, but it has other parameters, such as its speed, that can only be described numerically. For this purpose an *analog* representation of the value is used. Depending on the types of numbers that are used, an analog value can be represented as an integer or a fractional number with a decimal point.

Analog input signals take the form of changes in either voltage or current. The analog device may be measuring position, speed, flow, or another physical characteristic. These signals are connected to a circuit, which then converts the signal into a digital number. Output analog signals also take the form of changes in voltage or current. A digital set point is converted to an analog output, which may drive the speed of a motor or the position of a valve.

Analog inputs and outputs must go through these digital-to-analog and analog-to-digital conversions because of the inherently digital nature of computer and control systems. An analog value can have an infinite number of values within a given range. Pick any two points along the constant slope of a voltage change and there can always be another point between them.

Electrical signals are converted to digital from analog inputs using an analog-to-digital converter circuit (ADC). Signals are converted from digital to analog using a DAC, or digital-to-analog converter. These converter circuits are designed to operate over a fixed range of signals based on the application. The number of digital steps that an ADC or DAC is capable of is known as the *resolution* of the converter, this is described by the number of bits of the digital signal. A 16-bit DAC has a higher resolution than a 14-bit DAC, meaning it displays a higher number of subdivided values within its range.

Another specification related to analog signals is *linearity*. This is a definition relating to the "straightness" of the input signal or resultant conversion. Linearity may relate to aspects of the signal being measured or to the converting device itself. It may be thought of as how much the converted signal deviates from the original.

2.1.1 Scaling

Analog values must be converted into units of measurement to be displayed. The formula for doing this is derived from the formula for a line, $Y = m \times x + b$, where m is the scalar created by dividing the engineering unit range by the current or voltage range (also commonly referred to as the *slope* of the line), x is the analog value from the input point, and b is the offset (if there is any). Y is the engineering unit value to be displayed.

As an example, let us say we have a 4 to 20 mA input representing a weight in pounds. At 4 mA we have a reading of 0 lb, while at 20 mA we have a reading of 100 lb. Assume a 16-bit card that gives a reading of 0 at 4 mA and 65,536 at 20 mA. The range for pounds is then 100 and the range for the current is 65,536; the scalar is then $100/65{,}536 = 0.0015259$, which is the number of pounds per digital count. In this example, let us assume we have a value of 27,000 from our card. Multiplying by the scalar gives a value of 41.199, or about 42 lb. Note that in this example, there was no offset since both ranges started at 0.

Now let us use an example that does have an offset. Assume we wanted to know at what current value the weight of 20 lb would be. The scalar would then be $16/100 = 0.16$ mA/lb. Since the reading of 0 lb is at 4 mA, we have to use the offset b. The formula would then be $(0.16 \times 20) + 4$ or 7.2 mA. Another convenient way to get an approximate value is to simply graph a line on a piece of graph paper using the appropriate scales. Figure 2.1 illustrates how this is done for the above example.

Drawing a graph is also a great way to roughly check your math!

This process is even simpler when converting an analog signal to engineering units in a control program. Simply take whatever value is present when the process is at 0 (in the above example 0 lb) and subtract it from the signal. This is your offset. Then take the range from your new 0 to some known value (such as the 20 lb above) and determine the scalar: 20 lb/number of counts = scalar. As in the above example, this should be approximately 0.001526. This process can be automated to self-calibrate by using the resting value or unloaded value of the device to log the offset automatically and by using a calibrated weight to determine the range or scalar.

Important Concepts 11

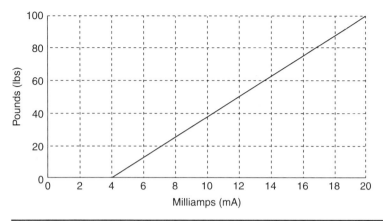

FIGURE 2.1 Analog scaling.

2.2 Input and Output (Data)

The control of a system reacts to input information and configures output(s) accordingly. Input and output information can be in the form of physical signals, such as electrical and pneumatic pulses or levels, or it can be in a virtual form, such as text instructions or data. A controller may react to switches or fluid levels by turning on valves or running motors at a given speed, or a computer may react to text or mouse-click-type instructions by changing display screens or running a program. These are both cause and effect illustrations of automation at work.

2.2.1 Discrete I/O

Most control systems on a manufacturing plant floor use discrete I/O (or input/output) in some form on both the input and output sides of the process. Digital signals, such as switches, push buttons, and various types of sensors, are wired to the inputs of a system. Outputs can drive motors or valves by turning them off and on.

Typical electrical I/O uses low-voltage and current signals for inputs and outputs. The most widely used signals are 24 volt (V) DC (24VDC) and 120VAC, although this can vary by application and by country. In some systems that need even lower electrical energy because of a hazardous environment, low-voltage systems called "intrinsically safe" circuits are used. These are typically about 8VDC or less. When a system is shielded from outside effects, such as signals inside a controller or on a circuit board, signals of 5VDC or less are common.

Because of personnel safety concerns, 120VAC I/O is not as widely used; however, systems that have sensors and actuators spread over a large physical area still sometimes use AC. Many older automation systems still use 120VAC, but 24VDC is more widely accepted on newer systems as electrical code requirements have limited the access to electrical systems over 60 V (refer to NFPA 70E). Process plants with AC valves and motor starters or large conveying systems still occasionally use 120VAC, but distributed communication or network-based I/O is becoming more common.

Other types of discrete I/O are used for special cases. Pneumatic valves can be plumbed in a configuration called "air logic," where switches may allow air to flow in a circuit, actuating valves and other air switches to serve much the same purpose as electrical signals. Air logic is used in some cases where electricity can be hazardous, but it is not as common as the use of electrical signals.

2.2.2 Analog I/O

Analog inputs and outputs typically take the form of changes in either voltage or current. Analog inputs may represent the position of a device, an air pressure, the weight of an object, or any other physical property that can be represented numerically. Most measurement systems use analog inputs. Analog outputs may be used to control the speed of a motor, the temperature of an oven, and many other properties. An example of how discrete and analog I/O signals differ is shown in Fig. 2.2.

Common analog ranges in industrial applications are 0 to 20 mA or 4 to 20 mA when using current or 0 to 10VDC for voltage. Current control is considered to be less susceptible to electrical noise—and hence more stable—while voltage control can be used over longer distances.

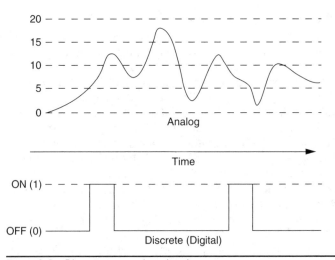

FIGURE 2.2 Discrete vs. analog signals.

PID Control

Control of a closed-loop system (Fig. 2.3) is often done with PID control algorithms or controllers. A closed-loop system takes feedback from whatever variable is being controlled, such as temperature or speed, and uses it to attempt to maintain a set point. PID stands for proportional-integral-derivative, the names of the variables set in the controlling algorithm. Another name for this is "three-term control."

In a closed-loop system, a sensor is used to monitor the process variable of the system. This may be the speed of a motor, the pressure or flow of a liquid, the temperature of a process, or any variable that needs to be controlled. This value is then digitized into a numerical value scaled to the engineering units of what is being measured. The variable is then compared to the set point for the system; the difference between the set point and the process variable is the error or difference that must be minimized by the system. This value is "fed back" into the system to counteract the error. Figure 2.4 shows a graphic diagram of PID control.

For any error that must be compensated for, there is some actuator or value that must be controlled to offset the error. In the case of temperature, this might be a proportional valve that feeds hot water into a system or gas into a burner; for a motor it might be current to increase speed or torque. The current error within the system is

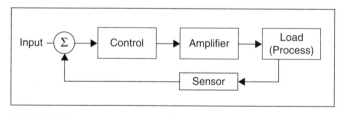

FIGURE 2.3 Closed-loop feedback diagram.

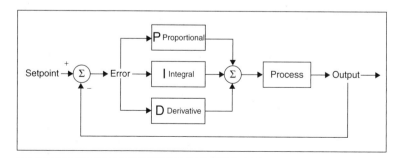

FIGURE 2.4 PID block diagram.

Chapter Two

closely related to the P or proportional value; in other words, the variable is used as a direct offset to the detected error.

One might think it would be sufficient to simply use the P value to constantly introduce an offset into a process; if one is trying to keep a container of liquid at a constant temperature, why cannot one just add heat until the container is at the desired temperature and then remove the heat? Experience would say that the temperature would either overshoot the set point or take a very long time to get there. There is the possibility that we would want to achieve the set point very quickly, further increasing the overshoot. This is where the other variables, the I and D parameters, are applied.

If the proportional variable is the current error, the integral or I value can be thought of as the accumulation of past errors, while the derivative or D value can be thought of as a prediction of future errors. These values are affected by the rate of change in the sensed PV and, if properly applied, can improve control of the process immensely. The I and D parameters are not always used in the process. One or the other is often omitted, creating the terms PI and PD control.

PID controllers may be a self-contained device such as a panel-mounted temperature controller or an algorithm within a PLC or DCS controlling an analog "loop." There are various ways to arrive at the P, I, and D values, including the Zeigler-Nichols method, "Good Gain" method, and Skogestad's method, but one of the most common is the "guess and check" or trial and error method.

Parameters and variables are set in an iterative process after defining the cycle. An example of this process for a temperature loop is illustrated here:

1. Predefine and set up a PV (process variable), CV (control variable), and SP (set point). In this example, assume PV is a temperature input 4 to 20 mA signal (RTD) downstream of the CV, where the CV is a modulating analog steam control valve. The set point would then be the desired water temperature the PV would need to achieve after steam is added.

2. Set integral and derivative variables to zero (0).

3. Start the process and adjust the proportional/gain tuning parameter until the PV starts modulating above and below the SP.

4. Time a cycle or period of this oscillation. Record this time as the natural period or cycle time.

5. After timing and recording the cycle, reduce the P (proportional) value to half of the setting needed to achieve the natural cycle.

6. At this point set the I (integral) parameter to the natural cycle. This will decrease the amount of time it takes for the PV to reach SP than with the P setting alone.

7. The D (derivative) may generally be safely set to approximately one-eighth of the integral setting. This value helps with "damping" or controlling the overshoot of the process. If a process is noisy or dynamics are fast enough, where PI is sufficient the D value may often be left at 0.

While moving from just the (P) setup to adding the (I) value, it is important to stop and restart the process as well as any setting adjustments that need to be made in a manual mode. If the process remains in automatic and attempts are made to adjust variables, the system will be looking at the previous cycles and will not be a true indication of PID performance in a typical start-up scenario.

There are many other possible modifications to hardware and software PID controllers, such as set point softening, rate before reset, proportional band, reset rate, velocity mode, parallel gains, and P-on-PV. Users should first select the standard PID form when setting up controller values. It is also important to make sure the loop update time is 5 to 10 times faster than the natural period.

Servo systems and software programs usually have auto-tuning algorithms that use results from input information, known system characteristics, and detected loads to approximate PID settings. Many manufacturers preset algorithm variables based on the hardware selected and input information about the process. Process control and servo system loops can act very differently, however, and a general knowledge of PID can be useful in setting initial parameters.

2.2.3 Communications

Communications methods can be applied to transfer larger amounts of information to and from a controller. With this method, digital and analog I/O statuses, along with text and numerical data, can be transferred. There are many different methods of communication-based input and output protocols. Many of the communication techniques described below have been adapted to allow remotely mounted devices and I/O blocks to be distributed to various locations on a machine or within a system and to be controlled from a central point. Often the remotely mounted I/O points can be semiautonomous in controlling their local stations, with only periodic communication to the central controller.

Devices and controllers are linked together to form a communications *network*. A network may be as simple as two devices talking to one another or a multilayered scheme with hundreds or even millions of devices on it (as with the Internet). Common topologies or layouts for networks include ring or star configurations (Fig. 2.5). An individual element of a network is also known as a *node*.

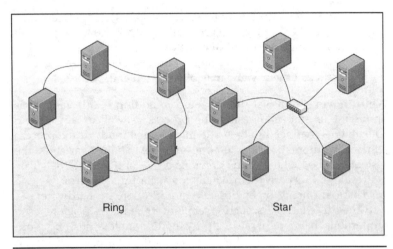

FIGURE 2.5 Ring and star topologies.

Serial

Serial communications are strings of digital 1s and 0s sent over a single wire. They can alternate between sending and receiving data or have a dedicated line for each signal. The protocols for the data sent across the lines can vary widely but a few of the common types of serial communications are RS232, RS422 and RS485. The RS in these designations is an acronym for "recommended standard" and does not describe the actual communication protocol being used.

RS232 communications typically use separate send and receive lines. These are labeled as TX for transmit and RX for receive. They can also use other lines such as CTS and CTR for clear-to- send and clear-to-receive as a traffic control or hardware handshaking method. There are a number of parameters such as baud rate (the speed of the communication or bit rate), number of bits per character (7 or 8), whether or not a "stop" bit is used, and whether the CTS and CTR lines are used (hardware handshaking). RS 232 signals are a common standard used between computer terminals and various manufacturers' control platforms. A nine-pin serial port is often included on a computer or control system, which makes it a convenient tool for downloading programming into control devices. The TX and RX lines can either connect to the same pin numbers on each end or connect RX-TX and TX-RX, which is known as a "null-modem" configuration. Adapters are readily available to reverse these pins or convert a male plug to a female one. These are commonly known as "gender benders."

RS422 and RS485 communications use a pair (or two) of twisted wires to carry the transmit and receive signals bidirectionally. Though it is not necessary to use twisted pair wiring, it helps with noise rejection. Many RS422 encoder cables do not use twisted pair because of their typically shorter length.

RS422 and RS485 can be used over a much longer distance and at a higher data rate than RS232 because it requires a lower signal voltage. RS422 is a multidrop configuration, whereas RS485 is a multipoint or "daisy-chain" configuration. These are often referred to as balanced or differential signaling (for example, 4 wire RS422 has RX+, RX−, TX+, and TX-). Over long distances, RS422/485 needs termination resistors at both ends (typically 120 ohm, like the CAN bus).

Serial ports also still exist in the Universal Serial Bus (USB) world, not only through USB/serial converters, but also because many USB devices use the USB port as a virtual serial port.

The protocols used in serial communication are typically based on strings of American Standard Code for Information Interchange (ASCII) characters. Information is generally text/alphanumeric based with line feeds (LF) and/or carriage returns (CR), indicating the end of a string of information. Device manufacturers often develop their own protocols for issuing commands or encoding data. This includes printers, test equipment, ID readers such as bar code and RFID, or simple text-based operator interfaces.

Parallel

Parallel communications allow multiple bits to be transmitted simultaneously over parallel lines. This can increase the throughput of data over RS232 signals, but it increases the cost of the cabling between two points. A common use of parallel cabling is between a computer's parallel port and a printer. Another common use of parallel communications is between CPU chips and the various registers used for processing data on a controller board. This configuration is easily visible when looking at the many parallel traces on a circuit board or the multicolored ribbon cables that often connect boards to each other. The backplanes of many control systems that connect controllers to their I/O cards are often parallel busses. Parallel communications are generally used over much shorter distances than serial communications.

Ethernet

Ethernet is a framework for computer networking technology that describes the wiring and signaling characteristics used in local area networks (LANs). The medium used for cabling Ethernet communications can be in the form of twisted pair wiring, coaxial cabling, or fiber-optic lines between points. As with the other communication methods described in this section, Ethernet only describes the physical characteristics of the system in terms of wiring and not the communication protocol used across the wires or fibers. Because of the widespread use of Ethernet in computing, nearly every computer is now equipped with an Ethernet port. Switches and hubs are used to connect computers and control devices in wide-ranging configurations. There are two different pin configurations for standard Ethernet cables: one with direct terminal-to-terminal configurations used with switches and hubs, and another known as a "crossover" cable for direct port-to-port connection.

Ethernet communications are very fast in comparison to serial and parallel communications and can transfer large amounts of data quickly. Devices are assigned a unique address at the factory called the MAC address, an abbreviation for media access control. This is a 48-bit binary address that is usually represented in hexadecimal with dashes, such as 12-3C-6F-0A-31-1B. Addresses must then be configured for each device on a network in the format of "xxx.xxx.xxx.xxx." Addresses can be configured directly by typing them into a field or configured automatically by a server using Dynamic Host Configuration Protocol (DHCP). Typical LAN systems use DHCP to prevent accidental duplication of addresses. A subnet mask is also used to prevent crosstalk between different connected networks.

TCP/IP is the set of communication protocols used for the Internet and other similar networks. This collection of standards is now commonly referred to as the Internet Protocol Suite. It gets its name from two of the most important protocols in it—the Transmission Control Protocol (TCP) and the Internet Protocol (IP), which were the first two networking protocols defined in the standard. Today's IP networking is a combination of several developments that began in the 1960s and 1970s. The Internet and LANs began to emerge in the 1980s and evolved with the advent of the World Wide Web in the early 1990s.

The Internet Protocol Suite may be viewed as a collection of wiring and signaling "layers." Each layer addresses a set of problems involving the transmission of data. Services are provided to the upper layers by the lower layers, which translate data into forms that can be transmitted. The stream of data being transmitted is divided into sections called *frames*. These frames contain both source and destination addresses along with the transmitted data and error-checking information. This allows the information to be retransmitted if it is detected to be different from what was sent originally. Errors are often caused by data "collisions" and require the data to be resent; this is more common as more devices are placed on a network. Because of this, the network speed is slowed dramatically and cannot always be reliably estimated. Standard Ethernet is considered to be a nondeterministic networking system because of this and is not suitable for direct I/O control.

Ethernet/IP is a subset of Ethernet often used in process control and other industrial control applications. Developed by Rockwell Automation and managed by ODVA, it is an application layer protocol, and it considers all of the devices on a network as a series of "objects." Ethernet/IP is built on the widely used CIP (Common

Industrial Protocol), which makes seamless access to objects from ControlNet and DeviceNet networks. Maximum response times can be configured and managed within this protocol, making it much more suitable for control applications.

A recent advancement in Ethernet topology is the Device Level Ring (DLR) which basically incorporates a two-port switch into each of the devices on the ring so that a level of redundancy can be achieved. Using this topology, even with the loss of communication between two devices, the entire system can still maintain communication. This topology does require a ring supervisor that determines how to send the packets of data to best manage the system and also monitor for any breaks in communication. Because of its wide acceptance in both IT and on the automation side, the Ethernet protocol seems to be a very widely accepted method of control and communication in newer automation topologies.

USB

USB is a configuration widely used in computer peripheral devices, but it is beginning to be adopted into automation systems. It was originally designed as a replacement for some of the RS232 and other serial connections on the backside of PCs. Along with communications between peripheral devices, it can also provide a limited amount of current to power devices. USB signals are transmitted on twisted pair data cable. Unlike some of the physical-only specifications described previously, the USB standard also includes frame and communications protocols for more commonality between devices from different manufacturers.

Special Automation Protocols

Many automation component vendors have developed their own protocols for communications using the various physical forms described above. Communication of data between controllers and operator interface touch screens are often developed by the manufacturer and, as such, are not used between different manufacturers. To facilitate communication between different manufacturers' devices, drivers are made available to allow devices to be easily interfaced.

Because of the interconnection problems between devices from different manufacturers, protocols have been adopted for communication and I/O control as standards.

Most of the following protocols are used for data communication and distributed I/O between a main controller and a remote node.

DeviceNet is an open communications protocol used to connect low-level devices, such as sensors and actuators, to higher-level devices, such as PLCs. DeviceNet leverages the controller area network (CAN) physical communication standard, which is a serial communication method for intelligent devices to talk to each other.

It uses CAN hardware to define a method of configuring, accessing, and controlling devices. It is commonly used for remote I/O and control of servo and other motor control systems. DeviceNet is not very fast compared to many other network methods, but it is considered very robust and reliable. DeviceNet networks can be difficult to implement initially and require special configuration software to commission.

CANOpen is a communication protocol used in embedded systems. It is also a device profile specification that defines an application layer for hardware. CANOpen consists of this application layer, an addressing scheme, and several smaller internal communication protocols. Because it is a mature, open protocol, CANOpen is supported extensively by servo and stepper controller manufacturers.

PROFIBUS is a bit-serial Fieldbus protocol developed by a group of companies in Germany. It is a global market leader among protocols because it can be used in both production automation and process automation. PROFIBUS PA is a low-current variation used to monitor measuring equipment in process automation (PA). PROFIBUS DP (Decentralized Peripherals) is used to operate sensors and actuators via a centralized controller in a production environment. Siemens is a major player in the PROFIBUS market.

Fieldbus is a group of industrial computer networking protocols developed for distributed control in real time. Prior to this development, computers were often connected using RS232 or other serial methods. Fieldbus includes most of the special protocols described above as well as Modbus, ASIbus, Sercos, ControlNet, and EtherCAT and other Ethernet-based networking methods such as EtherNet PowerLink, PROFINET, and Modbus/TCP.

In general terms, a *fieldbus* can be described as a network designed specifically for industrial control. This includes the protocols and physical layer definitions required to implement the communication system.

The highway addressable remote transducer (HART) protocol uses the 4 to 20 mA analog wiring of instrumentation devices as a medium for its communication signal. It superimposes a low-level digital signal over the analog value. HART technology is a master/slave protocol, which means that the slave device (the sensor or actuator) only replies when spoken to by the master device (the controller).

Though originally developed by Rosemount as a proprietary method of communication for its own products, it was made an open protocol for other vendors in the mid-1980s. There is a very large installed base around the world, mostly in the process industry. One reason for this is its ease of retrofitting into an existing wiring scheme. At only 1200 bps, it is not a very fast method of communication, but it is considered to be simple and reliable.

The signal includes diagnostic and status information from the device as well as the measured variable. The 4 to 20 mA signal is still considered to be the primary measured value and is unchanged by the superimposed digital communications. Communications are permitted in either point-to-point or multidrop network configurations.

Media for these various protocols vary from simple twisted pair wiring to shielded multiconductor or coaxial cable. Fiber-optic media can also be used in Ethernet communication and generally have a longer transmittable range than the various forms of copper media. Permissible distances between devices will vary based on the voltage level of the protocol and the configuration and type of media used.

Wireless

Wireless network refers to any type of computer network that is not connected by cables of any kind. This method avoids the more costly process of routing cables into a building or as a connection between distant equipment locations. Wireless telecommunications networks are generally implemented and administered over radio waves. This implementation takes place at the physical level or layer of the network structure.

A wireless local area network (WLAN) links two or more devices over a short distance, usually providing a connection through an access point for Internet access. The use of spread-spectrum technologies may allow users to move around within a local coverage area and still remain connected to the network. This is especially handy for devices such as portable HMIs.

Products using the IEEE 802.11 WLAN standards are marketed under the Wi-Fi brand name. Fixed wireless technology implements point-to-point links between computers or networks at two distant locations, often using dedicated microwave or modulated laser light beams over line of sight paths. It is often used in systems such as AGVs (automated guided vehicles), which move products around within a factory. It is also used with some RFID systems.

2.2.4 Other Types of I/O

High-Speed Counters

High-speed counters receive pulses at a high rate of speed. These pulses may indicate the speed or position of a motor or other revolving or linear device. High-speed counters are typically used with encoders and have other features such as inputs for homing signals and interrupts. They may be in the form of a stand-alone instrument or a card in a PLC rack.

Decoders

Decoders receive analog rotary position information from resolvers to provide angular position and velocity for a rotating axis. The input is typically a sinusoidal waveform in the millivolt range. Like high-speed counters, there are usually other input and output points available associated with the axis.

2.3 Numbering Systems

Various systems are used for numerical representation in the automation world. Some systems are configured for ease of use by computer- or microprocessor-based systems, while others are geared more toward high precision or being easily read by human beings. Table 2.1 illustrates how different radix-based numbering systems relate to each other.

2.3.1 Binary and BOOL

The binary numbering system is a base-2 system where each digit may only be a 0 (or "off" state) or 1 (or "on" state). Computers use this system internally because of the gate or switch logic nature of computing systems. The 1s and 0s can be grouped in such a way as to easily be converted into other numbering systems.

Though not directly related to the binary numbering system, logical operations on strings of binary characters are called Boolean or bitwise operations. For this reason, the binary system is sometimes referred to as a Boolean system, though not entirely correctly.

2.3.2 Decimal

The system most people are used to is the decimal or base-10 system. This system has a radix of 10 and allows fractional numbers to be represented conveniently using a radix point for base 10 or decimal point. Decimal numbers are not as easily converted to and from binary numbers because its radix of 10 is not a power of 2 like the hexadecimal and octal numbering systems.

2.3.3 Hexadecimal and Octal

The hexadecimal system is a base-16 system. Its primary use is as an easily convertible representation of groups of binary digits. It uses 16 symbols, most often 0 to 9 for the first 10 digits and A to F for the values 10 to 15. Since each digit represents four binary digits, it serves as convenient shorthand for values in base 2. A hexadecimal number is sometimes referred to as a "nibble," which consists of 4 bits as shown in Table 2.1.

The octal numbering system is a base-8 system that uses the numerical values of 0 to 7. Like the hexadecimal system, octal is easily converted from binary since it is grouped as three binary digits (its radix is an even power of 2). The octal and hexadecimal numbering systems are sometimes used as an I/O numbering base.

Decimal	Hex	Octal	Binary
0	0	0	0000
1	1	1	0001
2	2	2	0010
3	3	3	0011
4	4	4	0100
5	5	5	0101
6	6	6	0110
7	7	7	0111
8	8	10	1000
9	9	11	1001
10	A	12	1010
11	B	13	1011
12	C	14	1100
13	D	15	1101
14	E	16	1110
15	F	17	1111
16	10	20	10000
17	11	21	10001
18	12	22	10010
19	13	23	10011
20	14	24	10100
21	15	25	10101
22	16	26	10110
23	17	27	10111
24	18	30	11000
25	19	31	11001
26	1A	32	11010
27	1B	33	11011
28	1C	34	11100
29	1D	35	11101
30	1E	36	11110
31	1F	37	11111

TABLE 2.1 Number Conversion

2.3.4 Floating Point and REAL

Noninteger numbers are represented as floating point or real numbers. These are usually represented using 32 bits, also known as single precision. Sixty-four-bit floating point numbers are known as double precision.

Floating point numbers allow the radix point (decimal point in base 10) to be variable depending on whether a very large or very small number is being represented. Because the radix point (decimal point or, more commonly in computers, binary point) can be placed anywhere relative to the significant digits of the number, floating point numbers can support a much wider range of values than fixed-point and integers.

Floating point representation is advantageous in that it can support a wider range of values. However, the floating format requires slightly more storage to encode the position of the radix point and is slightly less precise than fixed-point numbers.

2.3.5 Bytes and Words

Bits can be grouped for convenience into 8-bit "bytes" and 16-bit "words." These numerical structures are convenient for passing packets of information that include numbers and text characters. Bytes can be further subdivided into 4-bit "nibbles," which can then be used to represent hexadecimal values. Table 2.1 illustrates this relationship further. Thirty-two-bit double words or double integers (DINTs) are also commonly used as a grouping technique.

For multiple byte data types (see Sec. 6.1.3 for more on this) bytes may be ordered in most significant value or "big end" in the first data register (big endian) or least significant value in the lowest data register (little endian). It is important to know how the data is arranged and used in computer registers since the wrong byte ordering will produce incorrect math results and mixed up text displays.

2.3.6 ASCII

ASCII is a coding standard that can be used for representing words and text characters. It is implemented as a character-encoding scheme primarily in computers and communications equipment. ASCII includes definitions for 128 characters. Thirty-three are nonprinting, mostly obsolete control characters that affect how text is processed. Ninety-four are printable characters, and the space is considered an invisible graphic. An "extended" ASCII table is included in the appendix of this book.

ASCII values are often grouped into user readable arrays called *strings* in programming. These arrays are arranged in such a way as to be readable by the human eye on HMIs, computers, or printed pages.

2.4 Electrical Power

Most control systems are electrical in nature with a few exceptions. Electrical power can be divided into two categories: alternating current (AC) and direct current (DC). AC is generated by rotating the windings of a rotor inside the stationary windings of a stator. Power generation plants use this technique to generate power at a very high voltage and transfer it across long distances. AC current changes polarity or direction many times a second. This waveform is represented using a sinusoidal shape (see Fig. 2.8 for a three-phase illustration). At the point of use, electricity is stepped down in voltage using a transformer to a more usable voltage. Typically DC voltage is then converted from a lower AC voltage using a DC power supply. The DC power is then used for devices in an automation system.

2.4.1 Frequency

The frequency of an AC voltage is expressed in hertz, abbreviated Hz. This is a unit of measurement of cycles per second, which directly relates to the rotary method used to generate the AC voltage. The frequency power-generating systems use in the United States is 60 Hz, while 50 Hz is common over much of the rest of the world. Ship- and aircraft-based power-generating systems often use 400 Hz.

Frequencies are carefully controlled at the power stations of most countries; however, in some remote areas, the frequency of the power grid can vary somewhat. In most cases, automation and control equipment can handle a slight variation in frequency, but some devices, such as AC motors, are designed for use at a specific frequency. It is important to know the power frequency of the country a system is designed for since many of the devices, such as transformers, motors, and power supplies, are designed around a specific frequency. There are often switches or internal jumpers that must be set to the correct frequency for proper equipment operation.

2.4.2 Voltage, Current, and Resistance

Subatomic particles called electrons flowing through a conducting medium such as a wire constitute electricity. The amount of electrons flowing through the conductor is known as *current*, which is measured in amperes, or amps. The force or "pressure" being applied to the current is called *voltage*, measured in volts. It is sometimes convenient to think of electricity in terms of water flow; the number of gallons per period of time being similar to current, and the water pressure being similar to voltage. If this analogy is carried further, the amount of restriction of water flow, such as kinking a hose, is similar to *resistance* in electricity, which is measured in ohms.

If any two values of an electrical circuit are known, the third can be determined using a formula called Ohm's law. This states that

Voltage (V) = Current (I) × Resistance (R). Conversely, $I = V/R$ and $R = V/I$. The letter E is sometimes substituted for V in these equations.

Resistors

Components that are made for the purpose of providing a fixed amount of resistance to current flow are called *resistors*. These are commonly made of carbon, although metal film, wirewound around a core, and various foils are also used. Carbon resistors often have a color code printed on them to identify their resistance value in ohms and their tolerance or precision (Fig. 2.6). Resistors are also classified by how much power they can dissipate; as current flows through a resistor, heat is created, which can damage the resistor and cause it to fail. Most low-power electrical circuits use currents that are well below the rating of the resistor.

Resistors can also be made in such a way that their resistance is variable. A wirewound resistor can be tapped at different points along its windings to vary the resistance between the tap and endpoint of the resistor. A diagram of these variable resistors is shown in Fig. 2.7. The first arrangement is known as a *rheostat* and is used to control current flow; I1 is variable in the diagram. If both ends of the windings are used along with the tapped point, the arrangement is known as a *potentiometer*. The potentiometer arrangement is also known as a voltage divider; V2 is variable in the diagram. Note also that I1 = I2 + I3. Both rheostats and

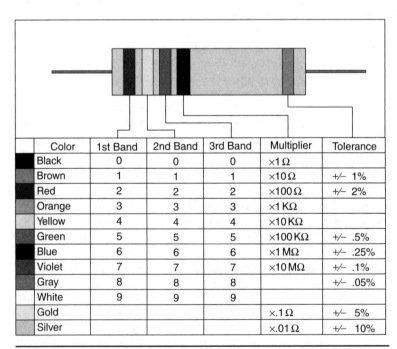

Color	1st Band	2nd Band	3rd Band	Multiplier	Tolerance
Black	0	0	0	×1 Ω	
Brown	1	1	1	×10 Ω	+/− 1%
Red	2	2	2	×100 Ω	+/− 2%
Orange	3	3	3	×1 KΩ	
Yellow	4	4	4	×10 KΩ	
Green	5	5	5	×100 KΩ	+/− .5%
Blue	6	6	6	×1 MΩ	+/− .25%
Violet	7	7	7	×10 MΩ	+/− .1%
Gray	8	8	8		+/− .05%
White	9	9	9		
Gold				×.1 Ω	+/− 5%
Silver				×.01 Ω	+/− 10%

FIGURE 2.6 Resistor color code.

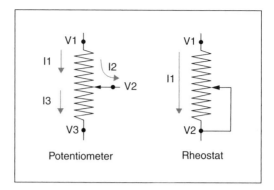

FIGURE 2.7
Potentiometer and rheostat.

potentiometers can be arranged in a linear fashion (slider) or a rotary configuration suitable for adjustment with a knob or dial.

2.4.3 Power

Power is the amount of electrical energy in a circuit or system and is expressed in watts. It can be calculated by multiplying voltage and current, or $P = V \times I$. Power is closely related to other units of energy, such as joules or horsepower, and can also be used to calculate the heat generated by a system or device if the efficiency is known. Power can be calculated in other ways for an AC voltage, depending on the number of phases of the electrical system. For larger power ratings, these values are usually expressed in kilowatts (thousands of watts), abbreviated kW.

2.4.4 Phase and Voltages

Electricity is usually supplied as a multiphase voltage. Three phases, each 120 electrical rotational degrees from each other, is very common in industrial facilities (Fig. 2.8). Though higher voltage is usually supplied at the service entrance of a building, it is generally stepped down using a transformer to various commonly used voltages. Common voltages for three-phase AC electrical power in the United States are 480, 240, and 208 V. Power is also often converted to single phase 120, 177, and 240 V. Voltages in some countries can be as high as 575VAC, so standard conductors are often rated at 600 V.

Voltages are not as precisely controlled as frequencies. A 480VAC voltage is usually measured at slightly below the stated voltage, while 220, 230, and 240VAC systems are often used interchangeably. Motors are specified for a range of voltages, as are DC power supplies.

Three-phase systems can be arranged in a delta configuration or a wye, which includes a neutral leg. The wye configuration produces not only the phase-to-phase voltage, but also an additional lower phase to neutral voltage. Additionally, windings can be center tapped to produce other voltages.

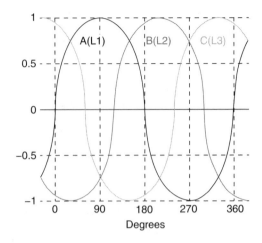

FIGURE 2.8
AC electrical phases.

2.4.5 Inductance and Capacitance

An *inductor* is a coil of wire. Any conductor creates a magnetic field proportional to the current flowing through it; by forming the conductor into a coil this magnetic field is concentrated into a smaller area. As the current is changed through the coil, the magnetic field tends to oppose this change. This property is known as inductance, and its unit of measure is the henry (typical inductor values are in the millihenry range). Inductors are used for filtering signals, shifting the phase of a voltage, or reducing noise in electrical circuits. Because of the opposition to current change, voltage is said to lead current when AC current is passed through an inductor.

Inductors may consist of multiple turns of wire with nothing in the center or may be wrapped around a ferrite (iron-based) core to increase the inductance. They may even take the form of a spiral etched into a circuit board for integrated circuits. A pair of inductors placed in parallel will induce a voltage from one coil to the other when alternating current is applied, forming a transformer.

A *capacitor* is formed of two plates of conducting material separated by an insulator, known as a dielectric. When a voltage is applied across the two plates, the capacitor builds up a static charge since the current cannot pass through the insulating dielectric material. This ability to hold an electrical charge is known as *capacitance* and is measured in farads (a standard capacitance value might be in the microfarads). If the voltage is alternated across the capacitor plates, the charges will tend to oppose the change. Capacitors therefore tend to have many complementary properties with inductors; where an inductor opposes a change in current, a capacitor opposes a change in voltage. An inductor passes current easily at low frequencies but increases its resistance to current as the frequency increases. This property is called inductive reactance and can be

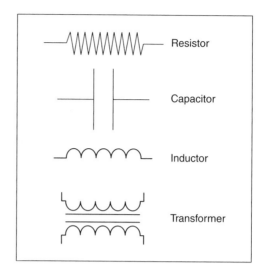

FIGURE 2.9
Passive electrical component symbols.

thought of as a variable resistance to AC. A capacitor passes current much easier as the applied frequency increases while creating a greater opposition to current flow at low frequencies, effectively becoming an open circuit when DC is applied (0 Hz). This frequency-based variation in resistance is called capacitive reactance. Current is said to lead voltage when AC current is passed through a capacitor.

Capacitors can take various forms; two plates placed in parallel with a thin piece of insulating material between them, long strips of conducting foil and insulating paper or plastic wound into a cylinder, or aluminum plate material separated by an oxidized material and an electrolytic liquid.

Historically, inductors were often called chokes, while capacitors were known as condensers, today the terms *inductor* and *capacitor* are most often used. Inductors, capacitors, and resistors are used together extensively in electrical circuits to control and filter signals and as supporting elements for solid-state circuits and integrated circuits (ICs).

These components are often termed passive components to differentiate them with active solid-state components (Fig. 2.9). Inductor and capacitors are also called "reactive" components because their response can vary depending on the frequency of the applied signal.

2.4.6 Solid-State Devices

Semiconductors are materials that have properties in between conductors and insulators. By introducing impurities into these materials, their properties can be changed to affect the electrical current that passes through them.

Current can be thought of as the flow of electrons through a conducting material in a direction determined by where electrons are repelled by a positive charge; hence, current flow is often indicated as being from positive to negative. Another way of describing current is known as hole flow; a space that an electron can fill can be thought of as a hole. Current flow can then also be thought of as the flow of holes from a negative to a positive direction. This can be useful when describing semiconductor devices.

Common semiconductor materials are silicon, germanium, and gallium arsenide, with the most commonly used being silicon based. The introduction of impurities into semiconductor materials is known as doping, which creates an excess of either electrons (N-type material) or holes (P-type material).

The simplest semiconductor device is called a *diode*. This is a device created by bonding a piece of P material and N material together. A characteristic of a diode is that when a voltage is applied across it, current can only flow in one direction. This makes it useful as a current control device in solid-state circuits. Diodes are commonly used in rectifier circuits that convert AC to DC. Certain types of diodes can also be used to detect light (photodiodes) or generate light (light-emitting diodes, or LEDs).

If semiconductor material is arranged is such a way that more than one P-N junction is formed, it is known as a *transistor*. A common type of transistor is created when two junctions are formed in an NPN or PNP arrangement. This is called a bipolar junction transistor and is often used in amplification applications. Other common types of transistor include JFET (junction field effect transistor) and MOSFET (metal-oxide semiconductor field effect transistor) types. Both of these make use of the property of semiconductors that an electrical field can change the conductivity of the material.

By applying a small voltage to one of the transistor leads, the current flowing through the device can be changed. When a transistor is placed in a circuit with constant resistances a small change in voltage on one lead can create a larger change in voltage at another; this is known as *amplification*. If the change in voltage is large enough a transistor acts as a switch, either allowing a large current to flow or nearly reducing it to 0. Amplifiers can therefore either be used in switching circuits or amplification circuits. Solid-state devices are considered "active" components because they can be used to convert a small low-voltage signal into a larger high-voltage or current signal. Figure 2.10 shows some of the schematic symbols used to represent active semiconductor devices.

Transistors are often wired together with other discrete components on printed circuit boards with metal traces and small holes for leads to be placed into. Components are usually attached by soldering and form transistor-transistor logic (TTL) circuits, amplifiers, or other types of electronic circuits. These can also be combined with IC devices to accomplish various electronic or processing functions.

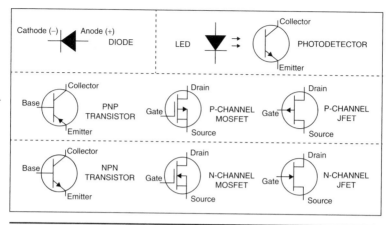

FIGURE 2.10 Active electrical component symbols.

TTL

Switching circuits can be combined into combinations that perform logic functions. An example might be producing an output signal only if certain input signals are present or not present. These circuits are often arranged as "gates" and are available in various combinations on integrated circuit chips. AND gates produce an output if *all* input signals are present or "high." OR gates produce an output if *any* input signal is high. The output of the circuit may be inverted (NAND or Not AND, NOR or Not OR) as may some or all of the input signals. This is usually designated by a small circle on the lead signifying an inversion. TTL arrangements can be arranged in vast combinations to form complex logical functions and form much of the basis of microprocessor logic.

These same logic symbols are often used in flowcharting for decision-making or logical processes along with other symbols. Figure 2.11 shows some of the standard logic symbols used in electrical or logic diagrams.

2.4.7 Integrated Circuits

When solid-state devices such as transistors and diodes are combined with passive devices such as capacitors, inductors, and resistors, circuits are formed that can alter electrical signals by amplifying, switching, and filtering them. These circuits may be made of discrete components mounted onto a circuit board or by patterning the components onto a semiconductor substrate material. These patterns form integrated circuits that can be miniaturized with millions of elements on a single semiconductor device. These devices may take the form of switching or amplification circuits, memory storage, microprocessors, or combinations of different types of circuits.

Integrated circuits are built up layer by layer on a silicon or other semiconductor wafer by imaging a pattern onto the surface, doping

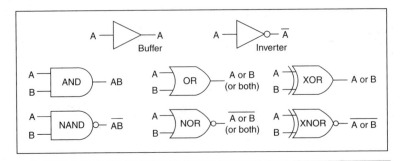

Figure 2.11 TTL symbols.

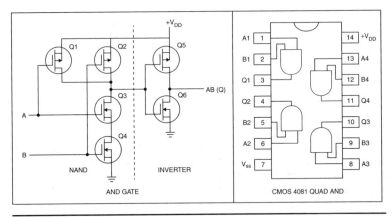

Figure 2.12 CMOS integrated circuit.

or depositing materials into the pattern, and chemically etching into the surface. Many layers can be formed in this fashion onto a single wafer, producing multiple "chips" or integrated circuits, which are then bonded to metal conductor leads to introduce signals or voltages into the chip. Figure 2.12 shows how transistors can be combined within a single chip to provide logic switching.

Complementary metal-oxide semiconductor (CMOS) technology is a common method of achieving high density for integrated circuits. CMOS circuits include image sensing, microprocessors, logic, data conversion, and communication transceiver circuits, among many others.

2.5 Pneumatics and Hydraulics

Collectively, the use of pressurized liquids or gases to drive actuators is known as *fluid power*.

Hydraulics define power generated by liquids, such as mineral oil or water, and pneumatics define power that is generated using gas, such as air or nitrogen. Fluid power can actuate cylinders in a linear fashion, motors in a rotational motion, or actuators in a rotational motion of less than 360°.

The advantages of hydraulics and pneumatics vary. While pneumatics are typically cheaper to implement and operate, they are less precise and are typically limited to larger utilities because of the high velocity of expansion once the gas is decompressed. On the other hand, while hydraulics might be more precise, they are more expensive to build and maintain because of their requirement for a means of liquid drainage and recovery and the need for larger equipment.

2.5.1 Pneumatics

Pneumatic or compressed air is used to move cylinders or actuators and is even used in some switching logic. Air is typically filtered, dried, and regulated to a usable pressure and distributed from a compressor to the various devices and actuators where it is required. A variety of quick disconnect devices and fittings have been designed in metric and standard sizes and are in wide use in industry. Pneumatic air supplies are distributed through most industrial facilities and are readily available for a wide range of uses.

Air pressure is usually applied to an additional filter regulator with a gauge for setting pressure at a machine or system. An additional lubricator is sometimes used to apply a minute amount of oil to lubricate the insides of air cylinders. Filter regulators (FR) and filter-regulator lubricators (FRL) are relatively low-cost and are regularly used at the entry point to a pneumatic control system. Additional pressure regulators, flow controls, and valves are used to control actuators for the desired effect. Control valves are common in 120VAC and 24VDC varieties. Figure 2.13 shows how these components are combined in a typical pneumatic circuit.

Air pressure in industrial facilities can be as high as several hundred psi, but typically pneumatic cylinders are operated at 60 to 80 psi. Pressure regulators can be used to reduce pressure within the system as required.

Flow controls can be placed on either or both ends of a pneumatic cylinder to regulate the speed of movement. The term *meter in* refers to restricting the inlet flow of air to a cylinder, while *meter out* refers to restricting the flow out of the other end. After a cylinder has been in position without airflow from either port, the pressure tends to slowly bleed off. This can cause a sudden "lunge" of the cylinder when pressure is reintroduced. To control movement in a selected direction, it is generally considered better to meter out rather than meter in to make the motion more consistent. In most pneumatic cylinders, flow controls are used on both ports and adjusted for desired movement in both directions.

Most valves are of the on/off or discrete variety, providing airflow (and therefore movement) to a cylinder until it is turned off.

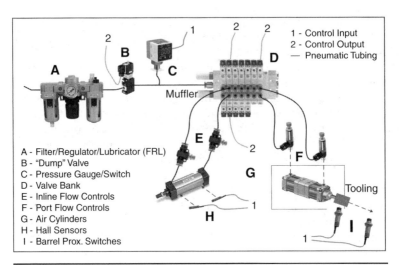

FIGURE 2.13 Pneumatic system.

Proportional air valves are also used to control motion by rapidly pulsing air or varying pressure or flow based on actuator position. This allows a simple form of motion control for position and speed.

The force that a cylinder can produce is based on the size of the bore of the cylinder and the air pressure available to the cylinder. The simple equation for the force a cylinder can produce is

$$\text{Force} = \text{pressure} \times \text{area} \tag{2.1}$$

Pressure in this equation is the available air pressure (in force/area units) and Area is the size of bore available to move the piston in the cylinder—typically Area = PI × bore radius^2 − rod diameter^2 (if the rod reduces the available area in the direction in which the force is being applied). This data is available in the specification sheets or catalogs from the manufacturer.

2.5.2 Hydraulics

Hydraulic systems, unlike pneumatic or air systems, typically have a pump located at each machine and are therefore more self-contained. Hydraulic pressure is used where more force is required than a pneumatic application. Like pneumatic systems, there is a wide variety of hydraulic cylinders and actuators available from manufacturers for each application. Hydraulic systems are much more costly to implement than pneumatic since the systems usually operate at higher pressure and more care must be taken to ensure that fluids do not leak from the system. Various other components, such as oil coolers and intensifiers, can also add to the cost of hydraulic systems.

Hydraulic systems do not operate as quickly as pneumatic systems, but, because of the lower compressibility of oil, they can be more precisely controlled. Pressures and force obtained from hydraulic systems are typically much higher than those of pneumatic systems and are therefore often used in metal forming applications.

2.5.3 Pneumatic-Hydraulic Comparison

Advantages of pneumatics:

- Cleanliness
- Simplicity of design and control
  o Machines are easily designed using standard cylinders and other components. Control is easy as it consists of simple on/off-type control.
- Reliability
  o Pneumatic systems tend to have long operating lives and require very little maintenance.
  o Because gas is compressible, the equipment is less likely to be damaged by shock. The gas in pneumatics absorbs excessive force, whereas the fluid of hydraulics directly transfers force.
- Storage
  o Compressed gas can be stored, allowing the use of machines when electrical power is lost.
- Safety
  o Very few fire hazards (compared to hydraulic oil).
  o Machines can be designed to be overload safe.
- Typically lower cost
  o Compressed air readily available in most manufacturing environments.

Advantages of hydraulics:

- Fluid does not absorb any of the supplied energy.
- Hydraulic systems operate at much higher pressures (typically 3000 psi) than pneumatic systems, providing greater force for a given bore size.
- The hydraulic working fluid is basically incompressible, leading to a minimum of spring action. When hydraulic fluid flow is stopped, the slightest motion of the load releases the pressure on the load; there is no need to "bleed off" pressurized air to release the pressure on the load.
- Can stop hydraulic cylinder in midstroke, where a pneumatic cylinder typically cannot.

2.6 Continuous, Synchronous, and Asynchronous Processes

Processes may take various forms in automated production. They may be continuous, as with the mixing of chemicals; synchronous, where operations are performed in unison; or asynchronous, where operations are performed independently. Manual and automated tasks may be mixed to utilize the decision-making and dexterity advantages of human labor.

2.6.1 Continuous Processes

Chemical, food, and beverage production often operates in a continuous fashion. Chemicals or ingredients are mixed together continuously to produce a "batch" of product. Plastics are often extruded continuously and then segmented into individual pieces for further operations.

2.6.2 Asynchronous Processes

Processes are said to be asynchronous when they do not rely on a master timing signal. An example of this might be an operation that takes place when a product arrives at an operator station from a previous process on a conveyor. The component may then be operated on when its arrival has been detected by a sensor rather than at the index completion signal from the conveyor.

2.6.3 Synchronous Processes

Synchronous processes rely on a master clock or timing signal. This may be an electrically or mechanically based system; cam-driven devices on a line shaft are examples of a synchronous process. Assembly-line operations may be synchronous or asynchronous, or a combination of both, depending on the source of the initiating trigger.

2.7 Documentation and File Formats

As automation systems are designed and built, documentation is generated both to convey fabrication information to the people building the system and to support it after it is in use. Computer-aided design (CAD) software, office software, and various proprietary packages are used to capture and plan the details of design. Specifications, requirements, documents, and manufacturers cut sheets are collected and usually presented as part of a maintenance manual for a machine or system.

There is a wide variety of file formats generated by the software that is used to create documentation. Many of these formats are proprietary to the companies that created the software, while some are open source. While Microsoft has dominated much of the office software arena since the PC has become common, picture or image formats have come from a variety of sources, each of which have their own advantages and disadvantages.

2.7.1 Drafting and CAD

Electrical and mechanical designs need to be conveyed to the people who will be fabricating parts and systems for automated machinery. Before computers, this had to be done by hand, using pencil and paper. Tools such as squares and triangles, protractors and compasses, and templates for devices and lettering were used to aid designers. Many engineers and draftsmen developed excellent lettering skills during their training.

Three-dimensional or isometric drawings as well as drawings of a part from three different views were often used for mechanical components. These drawings were often quite large, created on drafting boards with sheets of paper taped to them. They were then rolled up and placed in tubes for transportation and storage. Dimensioning of these mechanical parts was either in standard English, metric scales, or degrees, or used various drafting-specific symbols.

Electrical and logic drawings were done in the same way. Symbols for electrical or logical components were cut out of template material for tracing onto the page. Many of the symbols in use today are electronic versions of these template symbols.

CAD

As computers developed and became more economical, CAD was developed. This made design much faster and allowed engineers to do their own drawing rather than relying on draftsmen. Universities now offer classes in various software CAD packages, such as Pro/ENGINEER, SolidWorks, or AutoCAD, rather than drafting.

CAD software packages are either two-dimensional vector-based drafting systems or three-dimensional solid and surface modelers. Three-dimensional packages allow rotation in all dimensions and adjustment of color and transparency of various components. Viewers for assemblies are often available for free from the software designer so that customers can see designs without having to purchase the software.

Some CAD software is also capable of doing mathematical modeling of components. This allows strength and dynamic analysis of assemblies. It can then be used throughout the engineering process from the conceptual design or quoting stage through definition of the manufacturing methods for components. These dynamic modeling packages are often marketed as CADD for computer-aided design and drafting. Three-dimensional CAD is also known as *solid modeling*.

Electrical drawings and some mechanical design is still done using two-dimensional CAD. The most popular versions are made by Autodesk and marketed as AutoCAD. A less expensive version called AutoSketch is also available. File formats for AutoCAD end with the extension ".dwg," but drawings may be saved as .dxf files for easy exchange with other CAD programs.

Electrical drawings are referred to as *schematics*. Special symbols for electrical components are often kept in software "libraries" for repetitive use. Modern CAD systems can use these components to generate bills of materials (BOMS) for purchasing and spare parts. These systems can also analyze drawings to determine numbers of terminal blocks and labels, also generating off-page connector and line numbers for component labeling.

CAD drawings are typically printed or plotted to standardized page sizes. The most common for electrical drawing packages are D size, or 11-by-17 in. Electrical drawing pages usually have line numbers down the edge of each page, indexed by page number. This allows wire numbers to coincide with the line numbers for easy reference. Electrical drawings also have a title block, generally in the lower-right corner of the page border. This block contains project and page titles, revision numbers, dates, and initials for the people who drew, checked, and approved the drawings. This is an important part of the error-proofing process for electrical design.

Even with the checking process, mistakes and changes make drawing changes necessary after the drawings are released for assembly. Changes are often drawn in by hand using a red pen; this is known as "red-lining" a drawing. It is important that only one set of approved drawings be kept with red lines so that changes are kept for later drawing updates.

Mechanical drawings are often created in a three-dimensional software package but plotted to two-dimensional three-view drawings for fabrication. Like electrical drawings, changes may be red-lined by machinists or assembly technicians for later update. Mechanical drawings may be released on a wide variety of page sizes, depending on the complexity and size of the component or assembly. It is also common for a computer with a three-dimensional viewing software package installed for machine assembly to be present in the assembly area of a plant.

GD&T

Mechanical drawings are dimensioned using arrows and extension lines with tolerances in a fairly standard format, but for precision components or production parts additional information may be required. Geometric dimensioning and tolerancing (GD&T) uses a symbolic language on engineering drawings to define and communicate the nominal or theoretically perfect geometry of a part and its components along with the allowed variation. It also defines the possible variation in form and often describes the function of the part. It may describe the allowable variation in orientation and location between the

Important Concepts

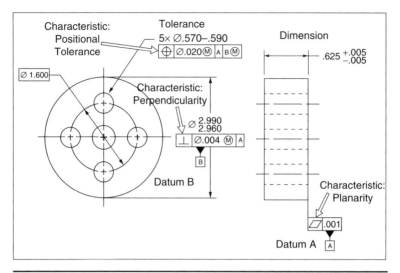

Figure 2.14 Geometric tolerancing.

features of a part. This is done using a datum reference, which is a theoretical ideal plane, line, point, or cylinder. A datum feature is a physical feature of the part identified by a datum feature symbol and corresponding datum feature triangle as shown in Fig. 2.14.

These are then referred to by one or more datum feature references, which indicate measurements should be made with respect to the corresponding datum feature and may be found in a datum reference frame.

The standard in the United States for GD&T is the American Society of Mechanical Engineers (ASME) standard Y14.5–2009. It provides a fairly complete set of standards within one document and is intended to describe the engineering intent for parts and assemblies. Other standards such as those issued by the International Organization for Standardization (ISO) vary slightly from Y14.5 and address one topic at a time. There are separate standards that provide the details for each major symbol and topic such as position, flatness, profile, and others.

GD&T is typically not used for design of automation machinery unless many machines must be built exactly the same way or precision manufacture of parts requires it. It is, however, used in the parts an automated machine may produce or assemble, so it is worth mentioning here. GD&T is used extensively in the automotive and production parts industries.

2.7.2 Other Design Packages and Standards

A variety of other documentation tools are used in specific industries, a few of which are described below.

P&ID

A piping and instrumentation diagram/drawing (P&ID) is a diagram used primarily in the process industry that shows the piping of the process flow together with the installed equipment and instrumentation. It is defined by the Institute of Instrumentation and Control as follows:

1. A diagram that shows the interconnection of process equipment and the instrumentation used to control the process. In the process industry, a standard set of symbols is used to prepare drawings of processes. The instrument symbols used in these drawings are generally based on International Society of Automation (ISA) Standard S5. 1.

2. The primary schematic drawing used for laying out a process control installation.

Figure 2.15 shows how a P&ID diagram for a simple spray booth might look.

P&IDs play an important role in the documentation, maintenance, and modification of the process that they describe. During the design stage, the diagram provides a basis for the development of system control schemes. It allows the visualization of the sequence of equipment and systems and their interconnections both electrically and mechanically. Though used primarily in the process and instrumentation fields, P&IDs can be useful as a method of communication between mechanical and electrical

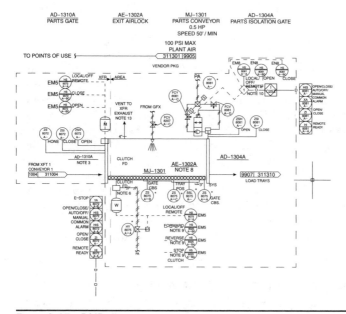

FIGURE 2.15 P&ID diagram.

designers when designing machinery. The diagrams also allow for further safety and operational investigations, such as for a hazard and operability study (HAZOP).

For processing facilities, it is a pictorial representation of key piping and instrumentation details, control and shutdown devices and schemes, compliance with safety and regulatory requirements, and basic start-up and operational information for a system.

Symbols and legends for P&ID devices can vary somewhat from company to company, but a legend with a key to the numbering systems, devices, and line styles is generally included in the drawing package. A few are shown in Table 2.2 for reference.

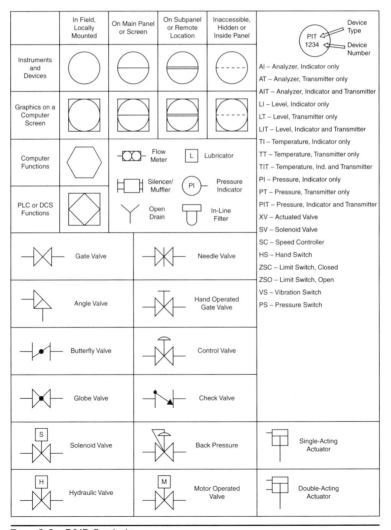

TABLE 2.2 P&ID Symbols

Adobe Acrobat and pdf Files

As software became more varied and new vendors appeared, methods were developed to convert files from one platform to another or present them in a common form. One of these is the Adobe Acrobat pdf reader and writer. This allows forms and documents to be printed to a reduced-size file that can be opened by anyone with a reader program installed on their computer.

The reader software can be downloaded for free off the Internet from the Adobe web site. Drivers to allow a file to be printed to a pdf document have also been created by a number of other developers. Typically these small programs are available for free or for a nominal charge after a free evaluation period.

A pdf writer is a bit more expensive and allows pdf files to be combined and modified from their native format. Since pdf files are similar to a picture of a document rather than the native document itself, it is not as easily formatted or changed once it has been converted outside text or other content. Pdfs have become one of the most common methods of publishing user manuals and other manufacturing documentation because of their widespread availability and smaller file size.

2.7.3 Image File Formats

As part of the documentation process, images or pictures are usually imported into documents. Image file formats are standardized means of organizing and storing digital images. There are a great many different types of image file used in documentation, but many can easily be converted from one into another by either shareware or vendor-specific software installed with printers or drawing packages.

Image files are composed of either vector-based geometric data converted to pixels by rasterization on a graphic display or files made up of the pixels themselves. An example of vector data is an AutoCAD drawing as described previously. The two main families of graphics are generally divided into raster (or pixel) and vector.

Pixels that constitute an image are ordered as a grid of rows and columns. Each pixel consists of numbers representing its grid location, magnitude of brightness and/or color, and possibly even elements such as transparency. The more numbers associated with each pixel, the larger an image file becomes. The greater the number of rows and columns becomes, the higher the image's resolution. Also each pixel increases in size as its color depth increases. An 8-bit or 1-byte pixel stores 256 colors or shades of black and white, while a 24-bit (3-byte) pixel stores 16 million colors. This is also known as *truecolor*.

Algorithms are often used to reduce the size of an image file. Higher-resolution cameras can produce huge image files of many megabytes. Not only may the image itself have a high resolution, but when using truecolor a 12-megapixel image can occupy 36,000,000 bytes of memory! Since a digital camera needs to store many images to be practical, various image file formats have been developed, primarily for reasons of image storage both within the camera and on a computer.

Other formats containing both raster and vector information also exist called *metafiles*. These files are application independent and are often used as a transfer medium from one application or software program to another. These are often known as intermediate formats, an example is the Windows metafile or .wmf, which can be opened by most Microsoft Windows–based applications. Another classification of formats that describes the layout of a printed page is "page description languages." Examples of this are PostScript, PCL, and the previously described pdf file from Adobe.

Including proprietary file types, there are hundreds of image file extensions. Following are just a few of the more common ones.

JPEG Files

JPEG or .jpg files are created by a compression method as described previously. The images are usually stored in the JPEG File Interchange Format, or JFIF. The algorithm produces relatively small files because it only supports 8 bits per color (red, green, and blue) for a 24-bit total. The resolution can also be reduced, and, as long as it is not by too much, the compression does not detract noticeably from the image quality. JPEG files do suffer from degradation when repeatedly edited and saved. Most digital cameras can save images in the .jpg format, many pdf generators also use it as their image compression algorithm.

TIFF Files

TIFF or .tif files are not as commonly used as JPEGs but are more flexible. They can be used to save files with either 8 or 16 bits per color in red, green, and blue for 24- and 48-bit pixels. They can be either "lossy" or lossless depending on the algorithm used. TIFFs can offer a good lossless compression for black-and-white (bi-level) images.

TIFF images are often used by optical character recognition (OCR) software packages to generate monochrome images for scanned text pages. Because of the flexibility of the format, there are few if any readers that can read every type of TIFF format that exists. Web browsers also do not support this format as commonly as with JPEGs.

PNG Files

The Portable Network Graphics (PNG or .png) file format is an excellent choice for editing pictures. It was created as a patent-free successor to the GIF file and is open source. Unlike the GIF format, it supports truecolor. Because of this, the file sizes are larger, but this format is optimum for images with large areas of the same color.

PNG is designed to work well online as it has a method of interlacing images that offers an early preview of an image even when only a small portion has been transmitted. It allows full streaming capability for online viewing applications like web browsers. It is also considered a very robust format since it checks for transmission errors and provides full file integrity.

GIF Files

The GIF or Graphics Interchange Format is commonly used for storing simpler graphics such as shapes, logos, and diagrams. The files are small because colors are limited to an 8-bit or 256-color palette. The format is often used for image animation but is not as effective for detailed images. It is one of the most common compression methods along with the JPEG.

BMP Files

The Windows Bitmap (BMP) or .bmp file is an uncompressed format that handles graphics within Microsoft's operating system. Because they are very large since they are not compressed, they are often converted to other formats for storage, but they are useful since they can be edited at the pixel level. They are often used within HMI programming software for graphic objects.

Vision systems and smart cameras also do not typically compress their image files so the files can be saved and analyzed as bitmaps. Since features are often quite small and may be analyzed to a sub-pixel level when measuring, machine vision images are generally left uncompressed, making this an excellent format for this purpose. Figure 2.16 shows a bitmap obtained from a vision system inspecting parts on a conveyor. The center part has been captured and processed by the camera to highlight and count the holes in the component.

2.8 Safety

When designing machinery for automated processes, one of the most important factors to consider is the safety of the personnel who will be using the equipment. Of secondary importance is the protection of the machinery itself. Because of the movement of machine components, hot surfaces, caustic substances, and sharp edges all pose potential dangers to exposed personnel. Because of this, many standards and regulations have been established as guidelines for the design of safety systems.

Each country and region has its own standards concerning safety, Restriction of Hazardous Substances (RoHS) and RF emissions. Though the European Union (EU) directives only specifically apply to member states, since the European countries are a major customer of most

Important Concepts 45

FIGURE 2.16 Machine vision bitmap.

companies, these companies find it necessary to comply with European regulations by obtaining CE marking, indicating compliance.

In 1995 a law was passed applying to all machinery built for use in the EU and European Economic Area (EEA). The law mandates that machine builders comply with the Machinery Directive on safety and indicate compliance by placing CE marking on their machinery. CE stands for Communate' Europe'ene, or European Community. Although this directive is European in origin, since components are sourced from around the world and the final destination of a product may not be known to the manufacturer, these regulations impact companies everywhere.

The Machinery Directive applies to machines functioning as a whole unit as well as interchangeable equipment modifying the function of a machine. It is a comprehensive set of rules applying to safety and is officially referenced as 89/392/EEC. These requirements cover every aspect of the machine: mechanical design, electrical design, controls, safety, and the potential for the machine to create hazardous situations. While the directive discusses controls and safety components, it only does so in the context of designing a safe machine as a whole.

The directive states that most manufacturers can self-certify, clearly stating exceptions. Some confusion arises when designers assume that this means if they are using components with CE marking, then their machines will meet CE requirements; this is not true. CE marking on a control component usually indicates compliance with the Low Voltage or Electromagnetic Compatibility Directive; this is an entirely different set of regulations. A similar analogy would be assuming that just because UL-listed components are used in a control panel that the panel meets UL requirements. To create a UL-approved panel, components must be wired and installed using an acceptable methodology, the National Electric Code. CE marking of machinery follows a similar reasoning.

Self-certification of equipment involves the examination of standards (types A, B, and C; applies to various aspects of safety by classification of machinery). Machinery is then tested against the requirements and "information for use" documentation is generated. If self-certification is allowed, a Declaration of Conformity is signed and the CE marking is affixed to the equipment. If self-certification is not permitted according to the Machinery Directive, an "EC Type Examination" is submitted for.

EN regulations or "European Norms" state the specific requirements of the directives. The key standards for clarification are EN-292–1 and EN-292–2, Safety of Machinery. EN-292 provides machine designers with basic concepts and terminology of machine safety, including safety critical functions, movable guarding, two hand control and trip devices, descriptions of hazards, and strategies for risk assessment and reduction.

The entire EC Machinery Directive (89/392/EEC) consists of 14 articles describing the requirements of manufacturers and European member states as well as procedures involved in the marking and associated documentation. There are also seven "annexes" or appendices elaborating on requirements.

China and Japan also have their own safety, RoHS, and RF emission and susceptibility standards. In many cases, if equipment meets the requirements of CE, it will also meet other countries' standards, but it is always a good idea to check the regulations of the country to which equipment is being shipped.

Whenever the potential for injury exists, it must be evaluated and safeguarded with at least the minimum required level of protection. In the United States, installation and use of machine safety and guarding is regulated by the Occupational Safety and Health Administration (OSHA). Some states also have their own safety organizations with regulations that must be at least as strict as the federal OSHA standards. A great reference for safety-related topics as they relate to the industrial workplace is the National Institute for Occupational Safety and Health (NIOSH) web pages on the Center for Disease Control (CDC) web site.

There are a variety of ways to safeguard operators from machine hazards. These include emergency stop (E-Stop) circuits, physical guarding, Lockout/Tagout, designed mitigation, guard devices, and software. In addition, several other agencies have established guidelines that influence safety design. The National Fire Protection Agency (NFPA) provides requirements for industrial systems. NFPA 79–07, the Electrical Standard for Industrial Machinery, is used for design of automated systems. The American National Standards Institute (ANSI) publishes the B11 standards to provide information on the construction, care, and use of machine tools. Since presses, cutting tools, and other processes are used in many machines and production lines, these standards often apply in addition to OSHA and NFPA regulations.

Standards in the B11 series include:

B11.1: Mechanical power presses

B11.2: Hydraulic power presses

B11.3: Power press brakes

B11.4: Shears

B11.5: Ironworkers

B11.6: Lathes

B11.7: Cold headers and cold formers

B11.8: Drilling, milling, and boring machines

B11.9: Grinding machines

B11.10: Metal sawing machines

B11.11: Gear-cutting machines

B11.12: Roll-forming and roll-bending machines

B11.13: Single- and multiple-spindle automatic bar and chucking machines

B11.14: Coil-slitting machines

B11.15: Pipe-, tube-, and shape-bending machines

B11.16: Metal powder compacting machines

B11.17: Horizontal hydraulic extrusion presses

B11.18: Machinery and machine systems for processing strip, sheet, or plate from coiled configuration

B11.19: Performance criteria for the design, construction, care, and operation of safeguarding when referenced by other B11 machine tool safety standards

B11.20: Manufacturing systems/cells

B11.19 is considered one of the best single sources of machine tool guarding information for the American market. This list also provides a good example of some of the many metal-forming techniques that have been automated and incorporated into machinery.

2.8.1 Hazard Analysis

To determine the level of risk to an operator or other personnel, a hazard analysis or risk assessment is performed. Classifications can then be made based on the results of the assessment and proper remedies applied. In most cases, there is more than one hazard present in a system; each must be addressed in the analysis.

In some cases, hazards can be completely eliminated by completely automating a process and removing human presence from the equation. This is not always possible because of cost or technology limits, however, and some risk may have to be accepted. To properly evaluate an application, the risk must be quantified. The potential consequences of an accident, the likelihood of avoidance, and the probability of occurrence must all be considered. Assessment of risk is then made by combining these in a matrix. Risks that fall into the "unacceptable" category must be mitigated by some means to reduce the level of safety risk.

The definitions below are from ANSI/RIA R15.06–1999, Table 1; Hazard Severity/Exposure/Avoidance Categories.

Severity:

S1—Slight Injury. Normally reversible, requires first aid as defined in OSHA 1904.12.

S2—Serious Injury. Normally irreversible or fatality. Requires more than first aid as defined in OSHA 1904.12.

Exposure:

E1—Infrequent Exposure. Exposure to the hazard less than once per day or shift.

E2—Frequent Exposure. Exposure to the hazard more than once per hour.

Avoidance:

A1—Likely. Can move out of the way, sufficient reaction time/warning, or robot speed less than 250mm/second.

A2—Not Likely. Cannot move out of the way, inadequate reaction time or robot speed greater than 250mm/second.

Severity, exposure, and avoidance are assigned a category according to the above criteria. A risk reduction category may then be calculated using Table 2.3.

This table is a combination of ANSI/RIA 15.06–1999 Tables 2 and 3. Using this chart, safeguards may then be chosen based on the results of the analysis. Certain task and hazard combinations, such as material-related tasks that include exposure to sharp parts, thermal, and ergonomic hazards, require the application of the highest level of feasible safeguarding and fall outside the scope of these tables. Appropriate standards and regulations should be consulted.

2.8.2 Emergency Stops

There are three different categories of stop functions: 0, 1, and 2. Category 0 is an uncontrolled stop by immediately removing power

			Risk Reduction Category	Safeguard Performance	Circuit Performance
S2 Serious Injury	E2	A2	R1 (Red)	Hazard elimination or Hazard substitution	Control Reliable Category 3
		A1	R2A (Red)		Control Reliable Category 3
	E1	A2	R2B (Yellow)	Engineering controls preventing access to the hazard or stopping the hazard. (e.g., interlocked barrier guards, light curtains, safety mats, or other presence sensing devices)	Single Channel with Monitoring Category 2
		A1	R2B (Yellow)		Single Channel Category 2
		A2	R2C (Yellow)		Single Channel Category 2
S1 Slight Injury	E2	A1	R3A (Green)	Noninterlocked barriers, clearance, procedures, and equipment	Single Channel Category 1
	E1	A2	R3B (Green)		Simple Category B
		A1	R4 (Green)	Awareness	Simple Category B

TABLE 2.3 Risk Reduction

to the machine actuators. Category 1 is a controlled stop with power to the machine actuators available to achieve the stop; then power is removed when the stop is achieved. Category 2 is a controlled stop with power left available to the machine actuators. The E-Stop circuit is a category 0 stop with only mechanical removal of circuit power allowed; that is, no software can be involved.

E-Stop buttons are an important safety component of most types of automated equipment. They are designed to allow an operator or bystander to stop the equipment quickly should anything go wrong.

E-Stop buttons are wired in series with a control circuit on a piece of equipment. Pressing the E-Stop button breaks the circuit and removes power from the holding relay that keeps the circuit energized. If the E-Stop device is a push button, it must be a mushroom-head-type and be red in color. There is also usually a yellow ring or background around the button. Figure 2.17 shows an E-Stop button on a control panel; note that the ring is not present in this illustration. This would not be acceptable for many safety systems; however, it often occurs on OEM equipment.

An E-Stop device does not have to be a push button, it may be a cable pull device or a guard door switch.

To energize the E-Stop control circuit, the operator presses the reset button. This provides power momentarily to a relay coil. The

FIGURE 2.17 E-Stop push button.

relay is often called a master control relay (MCR) or master safety relay (MSR). As the relay coil is energized, it closes its normally open contact so that power is provided through the relay contact to both the relay coil and the load. The load may be the power that feeds output cards or any other device that is hazardous. The power that is provided to the load is known as switched power.

As long as the relay is energized, the circuit is complete and power is provided to the load. Pressing the stop or E-stop button breaks the circuit, allowing the relay to return to its normally open position.

There are several degrees of reliability defined for E-stop circuits; simple, single channel, single channel monitored, and *control reliable*, which includes dual channel monitored circuits. Figure 2.18 illustrates a simple and a control reliable (monitored) circuit.

2.8.3 Physical Guarding

Physical guarding of a hazard is the simplest method of safety. A cover or other physical barrier is interposed between the hazard and the operator. The cover must be removed using a tool, or, if it is hinged as with a door, it must have a safety switch. Safety switches are available with locking mechanisms that are only released with an E-Stop condition if required by the safety and hazard analysis.

Mesh guarding is often used around machinery, allowing visibility into the machine. The size of the openings in the mesh is dependent on the distance from the hazard and the size of an operator appendage, such as a finger or arm. Figure 2.19 shows a mesh guard around an operating robot.

Hazard warnings and signage are required on machinery wherever hazards such as pinch points or electrical hazards exist. The symbols and colors for these signs are standardized and should not be created by individuals, but rather purchased from the appropriate vendor.

2.8.4 Lockout/Tagout

It should be noted that E-Stop systems cannot be used for energy isolation in a Hazardous Energy Control Procedure, or lockout. Devices for this purpose must physically separate the energy source from the downstream components. There are a number of requirements describing how branch fusing is applied, but all stand-alone machines must have a disconnect that removes power from the entire machine.

Disconnects are usually designed with a feature that allows a lockout device (usually a lock or a metal assembly that allows several locks to be inserted) to lock the disconnect in the deenergized position. A tag is then attached to each lock with the person's name who the lock belongs to and the date. Lockout/tagout is used whenever power needs to be removed from a machine or system for maintenance or when someone might be exposed to electrical power.

Chapter Two

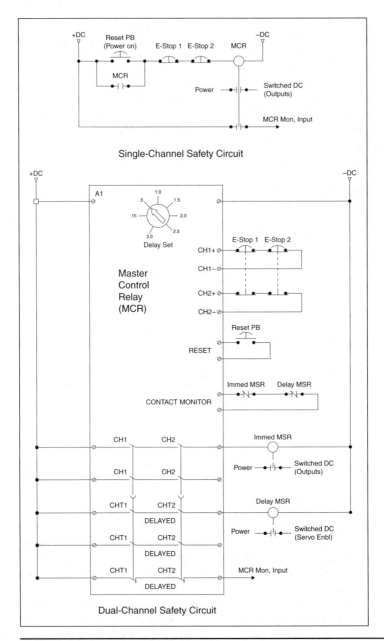

FIGURE 2.18 Single and dual channel E-Stop circuits.

Disconnects are often designed with a mechanism that prevents the opening of an electrical enclosure with the disconnect in the on position. Defeater mechanisms that allow the door to be opened with power applied using a tool are also often part of the disconnect design.

Important Concepts 53

FIGURE 2.19 Mesh guarding of a robot.

2.8.5 Design Mitigation

Another method of hazard reduction is by designing safety into the machine or system. An example might be rounding corners or placing moving parts and actuators in locations not easily accessible to personnel. This is usually the lowest-cost solution and a good designer will take this into account. Use of "finger-safe" terminal blocks and rubber bumpers or pads are examples of reducing exposure.

2.8.6 Guard Devices

There are many types of guard devices designed for safety applications. Some of the most common are light curtains and safety floor scanners. Nearly all of these devices are designed for use with control reliable (dual channel monitored) safety circuits. To accomplish this, circuit devices such as safety relay controllers monitor the actuation of the safety contacts with respect to the status of the received light beams. Light curtains are usually specified by the length of the covered area and the pitch or distance between beams. When specifying a light curtain, it is important to base the distance of

FIGURE 2.20
Light curtain.

the hazard to the guarding device on the reaction time of the safety circuit and its ability to stop motion before an operator can reach the hazard. There are software programs and testing mechanisms available to estimate appropriate distances and verify effectiveness of devices.

Figure 2.20 shows a 12-in quick disconnect light curtain.

Floor scanners and light curtains can also be "muted" to disregard areas of the curtain or covered area. This may be done on a permanent basis, such as teaching a floor scanner where fixed objects such as conveyor legs are, or conditionally, such as when an operator must enter a guarded area during normal operation. These types of circuits may also be automatically reset when the obstruction is removed. Typically this is done with a sensor such as a photo-eye sensing when an object is moving through the safety zone along a predefined path, such as an object on a conveyor.

Other types of safety devices are commonly used in guards circuits. Safety mats sense the weight of an operator and disable hazardous motions while the weight is present. Safety switches monitor the position of a door or gate that allows personnel access to a hazardous area. These types of sensors are often connected in a separate circuit from the E-Stops to allow an automatic reset. Circuit diagrams for guard circuits resemble those of safety circuits, removing power from output devices while personnel are located within the guarded area.

Important Concepts

A rapidly emerging component of safety systems is the safety controller. Whereas many safety devices and circuits must be hardwired and redundant, safety controllers allow logic functions to be included in the circuit without having to use relays and timers. Safety controllers may take the form of a "Safety PLC," where logic is programmed in ladder logic, or as a generic safety controller programmed by function selections and simple code. All safety controllers must meet the requirements of the Machinery Directive on safety, a European standard indicating compliance with CE.

Safety controllers provide a method of consolidating guard door switches, E-Stops, robot- and servo-enabled inputs into one control point. This allows circuits to be separated and prioritized while still meeting safety standards.

2.8.7 Software

In addition to the hardware methods described previously, additional steps must be taken to ensure software compliance with safety requirements. Hazard analyses for software are performed to identify potential hazards that may be caused by or prevented by programming. For software that controls or monitors potentially hazardous functions, the industry "best practices" for conducting a software safety hazard analysis are prescribed in IEEE STD-1228–1994, Software Safety Plans.

Like the ANSI hazard classifications described previously, the severity of the consequences establishes the criticality level of the software. Criticality levels are assigned from A to E; Catastrophic to No Safety Effect. The A and B levels require higher levels of rigor within the affected software task.

Analysis of software for risks must be performed by peers of the programmer or people knowledgeable of the programming language. Templates and programming standards also usually structure programs in such a way that conditions may be self-monitoring. An example might be ensuring that extended and retracted proximity sensors cannot be active at the same time or that a photo-eye is blocked only once during every cycle.

Built-in self-test (BIST) is often mandated for software programs. This may be a routine that checks an unattended machine to discover whether it needs maintenance or repair or perhaps determines whether a machine may be safely started. This is most common in military or computer applications but is sometimes seen in PLC and DCS programming.

Machine faults or alarms are created whenever safety devices are tripped or inappropriate conditions such as those described previously are detected. These faults may then deactivate or retract actuators to protect personnel, machine hardware, or products. Faults and alarms may trigger cycle or immediate stop commands or remove pneumatic force from a machine using an "air dump" valve or piloted check valves. In some cases, power may be removed from wired output terminals when a fault is detected.

2.8.8 Intrinsic Safety

When controls and instrumentation must be operated in hazardous environments such as explosive or flammable atmospheres, a protection technique known as intrinsic safety (IS) may be applied. In normal use, electrical equipment often creates internal tiny sparks in switches, motor brushes, connectors, and in other places. Such sparks can ignite flammable substances present in air. A device termed *intrinsically safe* is designed to not contain any components that produce sparks or that can hold enough energy to produce a spark of sufficient energy to cause an ignition.

Originally the IS concept was developed for process control instrumentation in hazardous areas such as gas platforms. The idea behind the application of IS is to ensure that the available thermal and electrical energy in an exposed environment is low enough that ignition of an explosive or flammable atmosphere cannot occur. IS barriers ensure that only low voltages and currents are present within the hazard area. Electrical signal and supply wires are protected with a Zener diode or galvanic isolation circuit within the barrier. Figure 2.21 is an example of several DIN-rail mountable IS barriers.

In the use of these barriers, a low-voltage (typically under 8VDC) device or sensor is connected to the field side of the barrier and the other side is wired into the standard inputs of a controller (typically 24VDC). The barriers are then located outside the hazardous area.

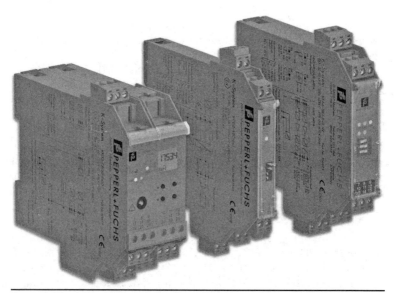

FIGURE 2.21 Intrinsically safe barriers. (*Courtesy of Pepperl+Fuchs.*)

Namur sensors are two wire sensors that change resistance when a target approaches the sensor. Namur is a German association that represents the users of measurement and control technology in the chemical industry. They standardized the use of intrinsically safe sensors typically used with barriers. When a Namur sensor "sees" a target, the resistance goes from low to high. This is opposite to the action of a conventional limit switch, where normally the resistance goes from high (infinity) to low (connected) when the switch is activated. This characteristic causes confusion frequently. They require an interface such as an intrinsically safe barrier that has a selection switch on the front that is used to reverse the action of the output relay contacts. This makes it behave like a limit switch (note that the action of the relay is reversed, not the action of the sensor). These sensors comply with European Standard EN 60947–5-6 (formerly EN 50227).

In the application of IS techniques, there are other considerations, such as abnormal component temperature. A short circuit inside a small component on a circuit board can cause the thermal energy within the immediate area to be much higher than during normal use. Current limiting by using resistors and fuses in branch circuits can be used to prevent autoignition of a combustible atmosphere.

Explosion-proof housings are also used to contain a fire or explosion within a confined area. These enclosures are usually sealed with multiple bolts through a removable cover, and special conduit is used to bring wiring to the device. These housings are often used in applications where non-IS devices or instrumentation must be used.

Chemical and petrochemical facilities are major users of IS devices and products. Grain processors and storage silos are also major users because of the flammable and explosive nature of organic dust.

Many sensor and device manufacturers have products designed for use in hazardous areas either with a barrier or as a self-contained unit. A well-known manufacturer of both intrinsically safe barriers and devices is the German company Pepperl+Fuchs. Further information can be found in their product literature.

2.9 Overall Equipment Effectiveness

Overall equipment effectiveness (OEE) is a method of monitoring and improving the effectiveness of manufacturing processes. Individual machines, manufacturing cells, and assembly lines all benefit from this lean manufacturing and total productive maintenance technique. OEE and productivity display screens are a common feature found on control system HMIs for machinery.

OEE takes the most common and important sources of manufacturing productivity loss, places them into three categories, and applies metrics to allow evaluation of the process. OEE evaluation and analysis begins with the amount of time that a production line is available for operation. This time is referred to as plant or line operating time. From this number a time category is subtracted called planned shutdown. This consists of all events that should be excluded from the OEE calculation because there is no intention of running the line. Examples of this include lunch and other breaks, scheduled maintenance, holidays, and periods of time when there is no product available to process. The time remaining after subtracting shutdown time from operating time is called planned production time.

In automated systems when a machine is in auto-run mode with no faults but is not running (often because of lack of operator-introduced material or operator stoppages), time is logged as "idle time." Machine builders often exclude this time from the downtime calculations during machine runoffs or factory acceptance tests (FATs) since they may be held to a minimum OEE for machine acceptance. Since operator-induced downtime is an important source of waste or "Muda," it should be included in the OEE calculation. Idle time may have to be discussed as a separate issue during runoffs if it is deemed to be excessive.

2.9.1 Availability

To determine the availability of equipment, efficiency and productivity losses are determined and analyzed. The first category of loss that is considered is downtime loss. This time is composed of any event that stops production for enough time to log an event. This can vary depending on whether events are logged automatically through the control system or by hand using an operator. Downtime includes such causes as equipment failure or faults, material shortages or "starvation," and changeover time. Faults and material shortages can be eliminated completely in an ideal situation, while changeover time should be minimized.

After subtracting downtime from planned production time, the remaining available time is called operating time. This can also be expressed as a percentage by dividing the operating time by the planned production time and multiplying by 100.

$$\text{Availability \%} = (\text{operating time}/\text{planned production time}) \times 100 \tag{2.2}$$

2.9.2 Performance

Machinery is designed to operate at an optimum speed. With no human involvement, this is usually fairly easy to calculate by evaluating a machine's operation under perfect conditions or looking at the speed of individual components within a system. Factors such as machine wear and tear, nonconforming product, operator inefficiency, and product misfeeds can all contribute to reducing a machine's designed ideal rate.

A machine's net operating time can be determined by subtracting downtime due to performance from the operating time that was determined previously. This can be difficult in an automated environment, so performance is often expressed as a percentage. A machine's or line's production output can be divided by its ideal output to determine a speed loss percentage. This can then be used to calculate downtime due to performance causes.

$$\text{Performance \%} = [(\text{ideal cycle time} \times \text{total pieces})/\text{operating time}] \times 100 \quad (2.3)$$

2.9.3 Quality

The quality of a process is determined by subtracting reject or defective parts from the total number of parts produced. The result can then be used to calculate a percentage of losses due to quality issues. This includes parts that have to be reworked.

Rejects can further be divided into losses that are incurred because of start-up activities and those that are truly bad parts. Activities like warm-up, splices, and setup errors should be differentiated from true failed parts. The goal of OEE calculation is to improve or eliminate causes of downtime and waste, so division of root causes into as many categories as possible can be useful.

$$\text{Quality \%} = (\text{good parts}/\text{total parts}) \times 100 \quad (2.4)$$

2.9.4 Calculating OEE

The results of the availability, performance, and quality calculations can be multiplied to determine overall machine OEE.

$$\text{OEE} = \text{availability} \times \text{performance} \times \text{quality} \quad (2.5)$$

Improving overall OEE is an important goal, but cannot be looked at so simply. For instance, few companies would trade a 5 percent improvement in availability for a 3 percent increase in rejects, even though the overall OEE would be better. All factors contributing to machine performance should be weighed and evaluated carefully.

Machine HMIs are often programmed with a productivity screen accessible from the main screen. Information such as shift times, ideal cycle times, and break/no-work times can be set by quality personnel, allowing OEE to be calculated and stored in the controller's memory. Data can then be recalled by shift, day, or week and evaluated along with downtime due to individual faults, manually logged operator events, and other data. Any cause that can be detected automatically should be programmed into the controls by category. Causes of downtime and reject types can be assigned "reason codes" for later evaluation.

2.10 Electrostatic Discharge

When two charged objects come into contact or close proximity with each other, a sudden flow of electricity can create a spark or voltage spike. This can be caused by a buildup of static electricity or by electrostatic induction. Electrostatic discharge (ESD) can cause damage to sensitive electronic devices, fires, or explosions.

In addition to precautions listed in the IS section of this chapter, extra steps must often be taken to prevent ESD-related hazards. For electronics manufacturers, even a small discharge that does not create a spark can be enough to damage a semiconductor device. When working with ESD-sensitive parts, typically everything within a specified radius of the part must be grounded. This area is known as an electrostatic preventive area (EPA). Parts within the EPA must be made of conductive materials such as stainless steel or electroless nickel-plated aluminum; certain dissipative plastics may also be used. Operators and technicians on the factory floor wear conductive-soled shoes, and the floor may be painted with conductive or dissipative epoxy paint. Conductive wrist and foot straps are often required, and electrostatic mats and humidity control are common. ESD prevention may extend outside the EPA by using ESD-safe packing materials and design techniques for external protection components or protection of input and output pins.

CHAPTER 3
Components and Hardware

Automated systems use a wide range of mechanical and electrical products from a great variety of manufacturers. Catalogs from the manufacturer, whether in paper or electronic form, can be a great resource for technical information not only on the manufacturer's specific product, but on general control and automation techniques.

3.1 Controllers

Controllers provide the computing, calculation, and I/O management part of an automation system. They may act as a nucleus or be networked together and distributed throughout the system.

3.1.1 Computers

In addition to being used as a tool to write the programs for control systems, computers are also used as the actual controller for some machines. Computers have the advantage of being relatively low cost because of their wide availability. Since computers already have a monitor and some sort of pointing device such as a mouse, human-machine interface (HMI) programs can also be easily implemented using standard computers.

Computer operating systems are not usually optimized for performing real-time control on machines. Most PC systems run a variety of the Microsoft Windows operating system, which by its nature contains many components not required or wanted in a control system. Because of this, a special platform, Microsoft Windows CE, was developed to remove many of the features not required for a control system. Windows CE is less memory intensive and component based and therefore more appropriate for real-time control. Embedded controllers are beginning to use Windows CE as a standard platform.

3.1.2 Distributed Control Systems (DCSs)

Distributed control systems (DCSs) are often found in process control applications such as chemical plants. They are used extensively in processes that are continuous or batch oriented. DCSs are connected to sensors and actuators and use set point control to control the flow of material through the plant. The most common example is a set point control loop containing a pressure sensor, controller, and control valve. Pressure or flow measurements are transmitted to the controller, usually through the aid of a signal conditioning I/O device. When the measured variable reaches a certain point, the controller instructs a valve or actuation device to open or close until the fluidic flow process reaches the desired set point. Large oil refineries have many thousands of I/O points and employ very large DCSs. Processes are not limited to fluidic flow through pipes, however, and can also include things like paper machines and their associated variable speed drives and motor control centers, cement kilns, mining operations, ore-processing facilities, and many others.

A typical DCS consists of functionally and/or geographically distributed digital controllers capable of executing from 1 to 256 or more regulatory control loops in one control box. The I/O devices can be integral with the controller or located remotely via a field network. Another name for this is distributed I/O. Today's controllers have extensive computational capabilities and, in addition to PID control, can generally perform logic and sequential control.

A DCS may employ one or several workstations and can be configured at the workstation or by an off-line personal computer. Local communication is handled by a control network with transmission over twisted pair, coaxial, or fiber-optic cable. A server and/or applications processor may be included in the system for extra computational, data collection, and reporting capability.

3.1.3 Programmable Logic Controllers (PLCs)

Programmable logic controllers (PLCs), are widely used to control plant floor automation systems. They are essentially digital computers used to control electromechanical processes. PLCs are used in many different industries and machines, such as packaging and semiconductor machines. Unlike general-purpose computers, the PLC is designed for multiple inputs and output arrangements, extended temperature ranges, immunity to electrical noise, and resistance to vibration and impact. Programs to control machine operation are typically stored in battery-backed or nonvolatile memory. A PLC is an example of a real-time system since output results must be produced in response to input conditions within a bounded time; otherwise unintended operation will result.

The main difference from other computers is that PLCs are armored for severe conditions (dust, moisture, heat, cold, and so on) and have the facility for extensive I/O arrangements. These connect the PLC to sensors and actuators. PLCs read limit switches or other sensors, analog process variables (such as temperature and pressure), and the positions of complex positioning systems. On the actuator side, PLCs operate electric motors, pneumatic or hydraulic cylinders, magnetic relays or solenoids, or analog outputs. The I/O arrangements may be built into a simple "brick"-style PLC, or the PLC may have digital and analog I/O modules that plug into a PLC rack. Rack-mounted communication modules can also be used to interface remote I/O blocks into the processor. An example of a rack-based PLC is shown in Fig. 3.1.

Major PLC manufacturers also sell the software to program their platforms. These software packages are specific to the platform; they cannot be used on other manufacturers' hardware. Additional software is often necessary to configure network communications and program HMIs; this may be packaged in a common software suite.

Prior to the advancement of computer systems, logic would be drawn manually using the same techniques as those used to design physical relay control systems and then converted into a shorthand that could be entered using a handheld keypad or text-based computer. As technology advanced, logic could be drawn on a computer screen. This was usually still converted into the same shorthand and available for documentation. Logic was often still printed out in a graphical format also.

Because of memory limitations, descriptive comments for coils, contacts, and other instructions were not stored in the PLC memory. Symbols could be used to create a reference for these devices, but generally they were simply referred to by bit or integer number. Memory registers were generally reserved for the type of data that was used; bit, word, or floating point values. Timers and counters also reserved areas as well as math registers.

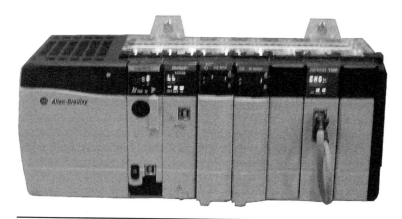

FIGURE 3.1 Allen-Bradley ControlLogix PLC.

In more modern PLCs, memory size is much less of an issue. More descriptive tags are allowed to be stored and other methods such as structured text and sequential function charts are able to be used. Programmers can even use a combination of these and ladder logic based on what is most appropriate. Further information on PLC software is contained in Chap. 6.

3.1.4 Embedded Controllers and Systems

Embedded systems are special-purpose computing systems that are usually designed to perform one or a few dedicated functions, often with real-time computing constraints. It is usually *embedded* as part of a complete device including hardware and mechanical parts. In contrast, a general-purpose computer, such as a personal computer, can do many different tasks depending on programming. Embedded systems control many of the common devices in use today. Common control components of an embedded system are the microprocessor or CPU, RAM or random-access memory, and flash memory.

In general, *embedded system* is not an exactly defined term, as many systems have some element of programmability. For example, handheld computers share some elements with embedded systems—such as the operating systems and microprocessors that power them—but are not truly embedded systems, because they allow different applications to be loaded and peripherals to be connected.

Temperature and Process Controllers

One type of special-purpose controller in wide use is the temperature controller. These devices may perform simple on/off heater control in reaction to a sensed temperature or control multiple zones using PID control. Controllers may be designed for a specific type of temperature sensor or configurable by software or dip switches.

Stand-alone temperature controllers are often sized using the DIN (Deutsches Institut für Normung) system, a German standard. They may thus be classified as 1/16 DIN, 1/8 DIN, or 1/4 DIN sizes. Timers and counters are also often sized this way. This ensures that controllers will fit a certain-size panel cutout. Figure 3.2 shows a 1/4 DIN panel mount temperature controller. Parameters such as the process value (PV) and set value (SV) are displayable in different colors for ease of use. Setting of parameters is done using the membrane keypad buttons on the front of the controller.

Temperature is not the only parameter that may be controlled by a separate controller. Nearly any process variable can be controlled using a stand-alone panel-mounted unit. Process controllers can be used to control the position of a valve based on flow or pressure; they effectively use an input variable to control an output variable as shown in Fig. 2.3, a closed-loop feedback system. Process controllers physically resemble temperature controllers; the major difference is the type of input circuitry.

Components and Hardware   65

FIGURE 3.2 1/4 DIN Temperature Controller. (*Courtesy of Omron.*)

3.2 Operator Interfaces

Machine operators need to interact with machinery in order to activate devices or processes and get status feedback. Historically this was done with push buttons, switches, and pilot lights. As technology has advanced, these have been replaced with dedicated text and graphic displays with membrane push buttons and touch screens. Industrial computers with a built-in monitor using a keyboard and pointing device such as a mouse are another form of machine interface. Computer interfaces with dedicated controllers are also widely used. The operator interface includes hardware (physical) and software (logical) components. Like the earlier description of automated systems, operator interfaces provide a means of:

- Input, allowing the user to send signals or data to a system or controller
- Output, allowing the system to control the effects of the users' manipulation

An operator interface is programmed by means of software on a standard computer terminal. The interface should be designed to produce a user interface that makes it simple and efficient to operate the machinery. The operator should be able to provide minimal input to produce the desired result, and the interface should provide only the desired information back to the operator. This requires careful planning of the screen menu structure and machine representative graphics and icons and consistently organized displays for an effective interface.

Other terms associated with operator interfaces are MMI (man-machine interface), HMI, GUI (graphical user interface), and OIT (operator interface terminal).

3.2.1 Text-Based Interfaces

Operator interfaces may be text based, simply providing instructions or machine status to the operator. They may or may not include buttons for input. The displays are typically backlit LCD but may be vacuum fluorescent bulbs. LED arrays are also common in larger message-only-type displays with a greater viewing distance. Individual lights are arranged in a pattern, allowing selected points to illuminate in the shape of alphanumeric characters. These may be arranged in multiple rows or columns, depending on the required message length and size of characters. Figure 3.3 shows production data displayed in a workcell. Colors are configurable within each field to make the display more readable; it can even be programmed for fields to change color based on numerical limits.

3.2.2 Graphical Interfaces

With improvements in technology, it has become standard for machines to use graphic interfaces with pictorial representations of the machine or production line for diagnostic purposes. These may be either monochrome or color and have membrane-type buttons, touch screens, or both.

FIGURE 3.3 Text display. (*Courtesy of Mills Products.*)

Components and Hardware   67

Graphic interfaces are manufactured by most PLC or DCS manufacturers and also by third parties who specialize in these products. They may use a proprietary operating system or be based on a computer platform, such as Microsoft Windows. Software for programming these operator interfaces is almost always proprietary to the manufacturer. Drivers are generally available for most popular controls platforms.

Graphical interfaces provide the ability to create a virtually unlimited number of screens and interface objects. Smaller screens can be superimposed over larger ones or minimized like the Windows operating system.

Faceplates may also be used with a graphical interface. A faceplate is an object that contains a standardized arrangement of buttons and indicators that may be populated via software with different device data. Thus if there are many similar devices, such as motors or conveyors, they can all use the same faceplate with their own start and stop buttons and status indicators.

3.2.3 Touch Screens

A touch screen is an electronic visual display that can detect the presence and location of a touch within the display area. Determining the location of a touch requires two measurement values, one on the X axis and one on the Y. The term generally refers to touching the display of the device with a finger or hand. It enables one to interact directly with what is displayed, rather than indirectly with a pointer controlled by a mouse, trackball, or touch pad.

Measurements coordinates are in analog form and are generally converted using a 10-bit analog to digital converter, providing 1024 positions in the X and Y directions. Touch points are then passed to a computer or HMI microprocessor using serial communication. Following are some of the different types of touch screen technologies available.

Resistive

Resistive touch screens are made of several layers of material. A hard outer surface provides insulation between an operator's finger and the inner conductive materials. Behind this layer are two thin electrically conductive layers separated by a small gap or space. The gap is separated by an array of very small transparent insulating dots; when the outer surface is pressed, the inner layers touch and the panel acts as a pair of voltage dividers. This creates electrical currents that indicate where the screen was pressed. This data is then sent to the controller for interpretation based on the address of the button or control that was drawn in that spot.

Resistive touch screens are cost-effective and are often used in restaurants and hospitals in addition to factories because of their resistance to liquids and other contaminants. A disadvantage of this technology is that it is easily damaged by sharp objects, such as tools. It also provides only about 75 percent optical transparency because of the extra layers and insulators.

Surface Acoustic Wave (SAW)

Touch screens using surface acoustic wave (SAW) technology use a glass surface and so are more resistant to sharp objects than resistive touch screens. SAW devices consist of two interdigital transducer arrays (IDTs) that transmit ultrasonic waves across the surface of the screen. Touching the screen surface absorbs a portion of the wave, registering the location on the surface. This is sent to the touch screen controller for interpretation.

Image clarity using SAW technology is better than that of resistive or capacitive touch screens because there are no extra layers between the image and the glass. Multiple touched points can also be sensed simultaneously. Contaminants on the surface of the screen can interfere with the ultrasonic waves, however. Because of the transmitting and receiving transducers being exposed along the edges of the screen, they are also not completely sealable and can be damaged by large amounts of liquids, dirt, or dust. They also must be touched with a fairly wide object, such as a finger; a hard stylus will not work.

Capacitive

A capacitive touch screen panel consists of an insulator such as glass coated with a transparent conductor. The human body is also an electrical conductor, so touching the surface of the screen results in a distortion of the screen's electrostatic field, measurable as a change in capacitance. The location is then sent to the controller for processing. Unlike a resistive touch screen, one cannot use a capacitive touch screen with an electrically insulating material, such as a standard glove. A special capacitive stylus or glove with fingertips that generate static electricity are sometimes used, but this can be inconvenient for everyday use.

There are several different capacitive technologies used for touch screens, each with its own cost or technical advantages. Surface capacitance technology provides a fairly durable and inexpensive product but has limited resolution, is prone to false triggering, and needs calibration during the manufacturing process. Projected capacitive touch (PCT) screens are more accurate because of a greater resolution. The top layer is also glass, making it more impervious to sharp objects. Because the conductive layer is etched, the clarity and light transmission are reduced, however.

Mutual capacitive touch (MCT) screens have a capacitor at the intersection of each row and column. This allows the registration of multiple touches to be detected, but these are more expensive than surface capacitive screens. Self-capacitive sensors can also be used with the same X-Y grid. They provide a stronger signal than the mutual capacitance type but cannot resolve more than one finger or touch at a time.

Infrared
An infrared touch screen uses an array of infrared LED and photodetector pairs around the edges of the screen to detect an interruption in the pattern of beams. These LED beams cross each other in a vertical and horizontal or X-Y pattern allowing the sensors pick up the exact location of the touch. This type of technology can detect nearly any input, including a finger, gloved finger, stylus, or pen. It is generally used in applications that cannot rely on a conductor (such as a bare finger) to activate the touch screen. Infrared touch screens do not require any patterning on the glass, which increases durability and optical clarity of the overall system, unlike capacitive or resistive technologies.

Optical Imaging
Optical imaging uses charge-coupled device (CCD) image sensors similar to a digital camera along with an infrared backlight. An object is then detected as a shadow. This can be used to detect both the location and the size of the touching object. As the cost of CCD components has lowered, this technology has become more popular. It is very versatile and scalable, especially for large-screen applications.

Dispersive Signal Technology
This technology uses sensors to detect the piezoelectricity generated in the glass due to a touch. Since the mechanical vibrations are used to detect the contact, any object can be used to touch the screen. Like SAW and optical imaging technologies, since there are no objects or etching behind the screen, the optical clarity is excellent. Because of the mechanical aspect of the technology, after the initial touch the system cannot detect a motionless finger.

Acoustic Pulse Recognition (APR)
Another interesting technology is acoustic pulse recognition (APR). Four small transducers along the edges of the screen detect the sound of an object hitting the glass. This sound is then compared using a lookup table to prerecorded sounds for every position on the glass. APR ignores ambient sounds since they do not match the stored digitized sounds. Like dispersive signal technology, after the initial touch the motionless finger cannot be detected, but the table lookup method is much simpler than the complex algorithm used to detect the piezoelectric contact.

3.3 Sensors

Sensors provide input data to control systems and can take many different forms. Discrete sensors may signal the absence or presence of an object or the position of an actuator, while analog sensors may be used to sense pressure, position, or many other physical qualities that can be described numerically.

3.3.1 Discrete Devices

Discrete sensors are digital in nature and provide an on or off signal. They often come with an attached cable for termination into a control cabinet, but they also have a variety of "quick disconnect" (QD) cabling options. They are typically available in 24VDC, 120VAC, or contact closure (relay) output configurations.

DC sensors use solid-state transistors as a switching method. There are two different types depending on the nature of input device they are being interfaced with: PNP, or "sourcing," and NPN, or "sinking." A sourcing sensor provides a positive reference signal to an input, or "sources" current. This means that it must be attached to a sinking, or NPN-type, input device. The opposite is true for a sinking sensor, which is connected to a sourcing input point that provides positive-to-negative current flow into the sensor.

QD cables are standardized for sensors; most are of the Micro or Pico QD variety. Cables are available in three, four, or five wire varieties, depending on the configuration of the sensor. For large devices such as light curtains, a QD cable with more conductors is specified.

Buttons, Switches, and Contact Closures

Buttons and switches are used by machine or system operators to signal a control system to perform a task or set a state, such as automatic or manual control mode. A push button typically has only two states, on or off, and may be maintained in each position (toggle)—momentary on or momentary off. Most push buttons are mechanical in nature and have a set of electrical contacts attached to the backside of the button. The contacts may be of the normally open (NO) or normally closed (NC) configuration. Some buttons may be touch sensitive or capacitive in nature with solid-state or mechanical contacts. Figure 3.4 shows schematic symbols for some of these different kinds of discrete input devices.

Selector switches may have multiple positions, each with a separate contact or group of contacts associated with it. Switches may be maintained at each position or spring return, giving the switch a "home" position or momentary effect.

Contact closures may also be controlled by the coil of a relay or solid-state signal. These are often wired to the inputs of a controller to indicate a status or condition. One by-product of using physical contacts in an electrical circuit is transients. Whenever a switch is opened or closed on an electrical circuit, a spike of voltage is created. Current is not interrupted immediately, and a small arc usually forms between the contact points. This can have an effect on the contacts themselves, causing pitting. It can also create a spark, which can cause problems in flammable or explosive atmospheres. If a controller

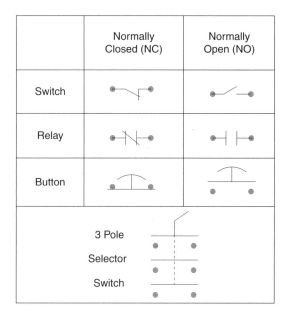

FIGURE 3.4
Contact closure schematic symbols.

is looking for a single change of state on an input, it can sometimes detect multiple "bounces" of the contacts due to transients. This can cause problems if an input is used as a counter.

Physical devices such as diodes are sometimes used on coils and contacts to help minimize these effects, while software "debouncing" can be used to ensure pulses are of a specified duration before accepting them as valid input. Solid-state devices are also commonly used to minimize the effects of transients.

Photo-Eyes

Photoelectric sensors, or photo-eyes, transmit and receive a light signal. The sensor changes state when the light changes from being received to not being received or vice versa. There are two conditions for the output of a photo-eye—Light On, where the output of the sensor is energized when light is detected, and Dark On, where the sensor output is energized when no light is being received. This is usually a selectable parameter with a switch or wire selection.

Photo-eyes are generally available in AC or DC varieties, although DC is much more common. DC photo-eye outputs are configured for PNP (sourcing) or NPN (sinking) outputs. There are also usually indicator lights on the body of the photo-eye for indication of power, switching status, or margin (amount of light received).

Usually, photo-eyes use an LED to generate the light signal. A lens is typically placed in front of both emitter and receiver to help amplify the light signal. The LED may be of various colors in the visible light range or in the infrared spectrum, which gives the light a longer range. Lasers are also often used for precise detection or longer-range applications. Visible LEDs are usually red, but green or blue are also used in diffuse or color-sensing applications.

When photo-eyes are placed too close together there is the possibility that light from one photo-eye's transmitter will trigger the receiver of a different eye. To reduce this possibility, manufacturers often modulate the light at different frequencies for different eyes within the same product group. Not all photo-eyes have this feature, but for those that do, there are typically series numbers or other markings that allow the eyes to be differentiated.

In addition to classifying photo-eyes by their output type, there are several physical configurations of the sensor.

Through-beam photo-eyes have a separate emitter, which transmits the light, and receiver, which receives the signal and controls the output state. Figure 3.5 shows this configuration. Because the sensor has to have two separate cables terminated for power and signal and it has two physical pieces, it is more expensive in both installation time and in hardware cost than other configurations. The through-beam photo-eye has a longer range and is considered to be most reliable when detecting the absence and presence of objects.

Retroreflective photo-eyes use a reflective tape or a plastic reflector to bounce the light off back into its receiver, as shown in Fig. 3.6. The emitter and receiver are both built into the same housing and use a common power wire, which reduces cost over the through-beam type, but the range is shorter.

To minimize interference from the light bouncing off other reflective surfaces, a polarized signal is often used. A corner cube-type reflector shifts the light 90° before it is received, and only light out of phase with the transmitted light is accepted as a signal. This allows the gain of the input circuit to be set at a higher level since the

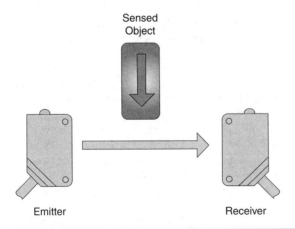

Figure 3.5 Through-beam photo-eye.

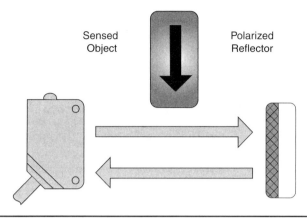

FIGURE 3.6 Retroreflective photo-eye.

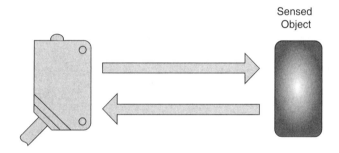

FIGURE 3.7 Diffuse photo-eye.

sensor will ignore signals from highly reflective objects that are not out of phase.

Diffuse reflective photo-eyes use the object to be detected as a target, as shown in Fig. 3.7. Light is transmitted from the emitter and received similar to the other configurations; however, the light received indicates the presence of a target rather than its absence, as with the through-beam and retroreflective configurations. Diffuse reflective photo-eyes are not the ideal sensor to use for simple absence or presence of an object since the amount of light received is affected by the reflectivity and color of the target. This property can, however, be used to an advantage when using the sensor to differentiate between colors. A red LED signal will reflect much more strongly from a red object than from a green object and vice versa. Techniques using red, green, blue, yellow, and white LEDs as light sources are often used in color detection photo-eyes.

There are also various additional physical configurations for photo-eyes. The amplifier that powers the transmitting and receiving

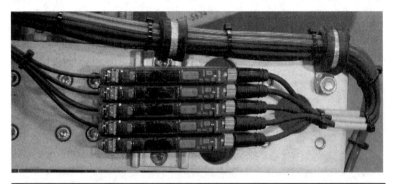

FIGURE 3.8 Fiber-optic amplifiers.

LEDs may be self-contained with lenses or it may have a head for attachment of fiber optics; see Fig. 3.8). Fibers may be made of plastic or glass. Plastic fibers usually have an opaque plastic cladding around the outside the clear inner fiber. This protects the fiber and acts as a waveguide for the light to reflect back into the core, keeping the light confined. Glass fibers typically can be used in longer-distance applications; however, they are more fragile than plastic fibers. Plastic fibers can be bent in a tighter radius than glass fibers.

Lenses and right-angle tips are often used for fiber optics. The mounting end of fiber-optic cables may also be threaded for use with nuts. Plastic fibers may be cut to length using a small cutter that is often available from the photo-eye manufacturer. Glass fibers usually have factory ends on them to prevent damage and must be purchased in the required length.

Proximity Switches

Proximity switches are used to detect the position of an object. While photo-eyes are also sometimes referred to as proximity switches, here we are discussing inductive, capacitive, limit switch, and Hall effect types. Proximity switches are often called proxes for brevity.

Inductive proximity switches are used to detect metal objects. A coil of thin wire is energized with a weak current that is connected to an oscillating circuit. When a large enough piece of metal enters the field created by the current flowing through the coil, the oscillator is stalled and a discrete signal is generated, signaling the presence of an object. The type of metal being detected strongly influences the range of an inductive proximity switch. Metals, such as steel containing iron, make the best target, while aluminum reduces the sensing range by about 60 percent.

In addition to the PNP or NPN output designation, inductive proximity switches are categorized as being shielded or unshielded. Shielded proxes have a metal housing all the way up to the sensing face of the switch. This reduces the range but allows the sensor to be

FIGURE 3.9 Inductive barrel proximity switch.

mounted flush with a metal surface without detecting the metal to the side. Unshielded proxes have a longer range since the field extends out from the sides of the prox, but are more susceptible to damage or interference.

Inductive proximity switches are available in a threaded barrel style, a flat surface mount, or various other configurations. Barrel proxes typically have a metal housing with a plastic-covered sensing surface, but they can be made entirely of stainless steel for ruggedness. Figure 3.9 shows a shielded barrel proximity switch sensing a metal target block. Note that it is threaded completely into its mounting block, indicating that it is shielded.

Capacitive proximity switches use a capacitive sensing surface that discharges when an object is placed close to it. As such, it can be used to detect nonmetal solid or liquid objects. A common use of capacitive proxes is to detect a liquid through the sides of a plastic or fiberglass vessel. As long as the vessel walls are fairly thin, the prox can be set to detect the difference in mass between an empty and full vessel. Capacitive proxes are also sometimes used as operator push buttons for ergonomic purposes since they take no pressure to activate, unlike mechanical push buttons.

Like the inductive proximity switch, capacitive proxes have a very short detection range. They are usually larger than their inductive cousins.

Hall effect sensors create a voltage difference based on the amount of magnetic field they sense. Hence, they are used to sense magnetic objects, such as a magnet moving with a piston inside a cylinder body.

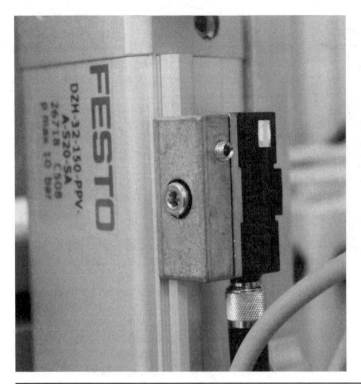

FIGURE 3.10 Hall effect sensor on pneumatic cylinder.

Any current-carrying conductor creates a slight magnetic field of its own, transverse to the direction of current flow. With a known magnetic field, its distance from the Hall plate can be determined. This makes Hall effect sensors a great choice for end-of-cylinder stroke sensors since they can sense the magnet through a metal (typically aluminum) body. This is illustrated in Fig. 3.10; inside the aluminum body of the Festo cylinder is a magnet attached to the cylinder's piston head. These electromagnetic transducers are used for proximity switching, positioning, speed detection, and current-sensing applications.

Hall sensors and inductive proximity switches are the most common sensors used in detecting cylinder or actuator position. What is the major difference between a Hall switch and an inductive prox? Essentially, a Hall effect sensor can sense a magnetic field, whereas an inductive sensor creates its own magnetic field.

Limit switches are mechanically activated devices that open or close electrical contacts when an object contacts it. There are wide variety of configurations, sizes, and degrees of ruggedness for limit switches. Roller limit switches have a metal or plastic roller that allows an object, such as a cam, to slide along the contact point. Lever arm and "whisker"-style switches extend the reach of the switch.

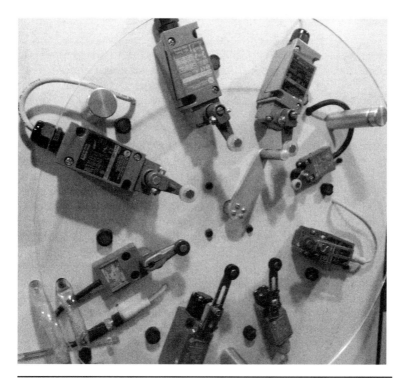

FIGURE 3.11 Roller limit switches.

Figure 3.11 shows several roller and lever arm switches mounted on a rotating display at a trade show; as the arm rotates in the center of the display, the lever arms move actuating the internal switch contacts.

Precision limit switches are used to precisely control the actuation point of a switch for positioning or measurement. They are typically plunger-type switches with a very short stroke.

3.3.2 Analog

Analog sensors produce an output that is proportional to a measured property. There can often be offsets and linear errors associated with analog sensors that must be taken into account when using the resulting measurements, and calibration to a known standard is often required. Analog sensors are often known as *transducers*.

Analog sensors are often used in automated and manual gauging. Special-purpose machines are often built around a specific type of gauge or group of gauging devices as a test station.

Pressure, Force, Flow, and Torque Sensing

Force can be measured using a variety of devices. One common element in measuring the amount of force exerted on an object is a *strain gauge*. Because strain gauge wires are fragile and difficult to

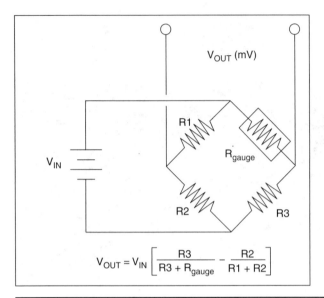

FIGURE 3.12 Wheatstone bridge strain gauge circuit.

handle, they are typically attached by an adhesive, such as superglue, to an insulating flexible backing, such as plastic. As stress is applied to the mounted strain gauge, the foil is deformed, causing its electrical resistance to change. The resulting resistance change—usually measured using a Wheatstone bridge circuit—is related to the strain by the "gauge factor," or the ratio of electrical resistance to mechanical strain, factoring in temperature, which also plays a small role. This circuit is illustrated in Fig. 3.12.

A strain gauge can be configured in a variety of physical packages to measure force or weight or to determine vibration and acceleration.

For small strain measurements, semiconductor strain gauges known as piezoresistors are preferred because they usually have larger gauge factors—or changes in resistance over a range of strain—than a foil gauge, thus allowing for more accuracy. Downfalls of semiconductor gauges include the higher cost, fragility, and greater sensitivity to temperature changes.

A *load cell* is a transducer that converts an input mechanical force into a measurable electrical output signal. When weight, or load, is applied, the strain gauge deforms, changing the electrical resistance of the gauges in proportion to the load. The strain gauge measures the deformation, or strain, as an electrical signal as current is passed through the gauge element.

In order to ensure maximum sensitivity and account for temperature changes, the typical load cell consists of four strain gauges in a Wheatstone bridge configuration. Load cells with one strain gauge, a quarter bridge, or two strain gauges, a half bridge, are also available.

Because of the small amount of electrical signal output produced, typically only in the range of a few millivolts, amplification by an instrumentation amplifier is required. The amplified output is then fed into an algorithm to calculate and scale the force applied to the transducer.

Pressure may be measured by using a piezoresistive strain gauge as described previously. The gauge is attached to a force collection element, such as a diaphragm, piston, or bellows, and deflection is measured proportional to the change in pressure. Absolute, differential, gauge, and vacuum pressures can be measured using this method. A diaphragm with a pressure cavity can be used to form a variable capacitor that is effective in detecting low pressure changes. Displacement of the diaphragm can also be measured inductively by measuring deflection of a magnet, use of a linear variable differential transducer (LVDT), or detection of an induced eddy current. These methods are known as electromagnetic pressure sensing. Optical methods can also be used by detecting changes in light transmission through an optical fiber as it is deformed.

Flow of liquids or gases can be measured in a number of ways. A rotary potentiometer (resistive element) is often used when attached to a vane that turns in the fluid or gas. Other flow sensors are based on devices that measure the transfer of heat caused by the moving medium. This principle is common when using microsensors to measure flow. Flow meters are related to devices called velocimeters that measure velocity of fluids flowing through them. Laser-based interferometry is often used for airflow measurement, but for liquids, it is often easier to use a physical deformation of some kind to measure the flow. Another approach is Doppler-based methods for flow measurement. Hall effect sensors may be used on a flapper valve, or vane, to sense the position of the vane, as displaced by fluid flow.

Detection of flow and pressure along with the measurement of valve positions in the process industry is known as *instrumentation*. Analyzers that detect properties such as acidity, viscosity, or density can also be included in this group. Outputs from instrumentation are often connected to *transmitters*, which convert signals into standard ranges such as 4 to 20 mA or 0 to 10 V signals.

Commonly, *torque sensors* or torque transducers use strain gauges applied to a rotating shaft or axle. Because of the relative movement of the shaft a noncontact means to power the strain gauge bridge is necessary, as well as a means to receive the signal from the rotating shaft. This can be accomplished using slip rings, wireless telemetry, or rotary transformers. Newer types of torque transducers add conditioning electronics and an ADC to the rotating shaft (rotor). Stator electronics then read the digital signals and convert those signals to a high-level analog output signal, such as +/−10VDC

Color and Reflectivity

As described in the digital sensor section 3.3.1, various colors of LED light reflect from different colored materials with varying intensity. This property can be used to sample the amount of light returning to a receiver and determine color. Combinations of reflected red, green, and blue light can be analyzed to determine shades and hues to separate items of different color properties. Despite this being listed in this current section, color sensors are often "taught" a color and an output is then switched if the color is detected.

For more accurate determination of color, a CCD is used to capture a colored region. CCDs react to photons, and when a filter called a Bayer Mask is placed over the CCD, it becomes a color-sensitive device. Red, blue, and green again are the operative colors for color CCDs. CCDs are also used to create black-and-white images that can be converted to a scale for intensity measurement.

LVDTs

LVDTs are a type of electrical sensor used for measuring linear displacement. The transformer-like device has three solenoidal coils placed end to end around a tube. The center coil is the primary, and the two outer coils are the secondaries. A cylindrical ferromagnetic core, attached to the object whose position is to be measured, slides along the axis of the tube.

An alternating current is driven through the primary, causing a voltage to be induced in each secondary proportional to its mutual inductance with the primary. The frequency is usually in the range 1 to 10 kHz.

As the core moves, these mutual inductances change, causing the voltages induced in the secondaries to also change. The coils are connected in reverse series, so that the output voltage is the difference (hence "differential") between the two secondary voltages. When the core is in its central position, equidistant between the two secondaries, equal but opposite voltages are induced in these two coils, so the output voltage is 0.

When the core is displaced in one direction, the voltage in one coil increases as the other decreases. This causes the output voltage to increase from 0 to a maximum. The output voltage is in phase with the primary voltage. When the core moves in the other direction, the output voltage also increases from 0 to a maximum, but its phase is opposite to that of the primary. The magnitude of the output voltage is proportional to the distance moved by the core (up to its limit of travel), which is why the device is described as "linear." The phase of the voltage indicates the direction of the displacement. Figure 3.13 illustrates the internal arrangement of an LVDT.

Because the sliding core does not touch the inside of the tube, it can move without friction, making the LVDT a highly reliable device.

Components and Hardware 81

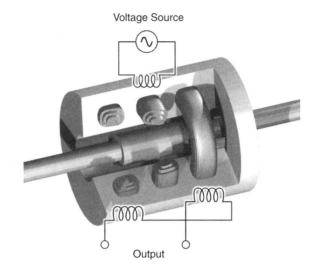

FIGURE 3.13
LVDT.

The absence of any sliding or rotating contacts allows the LVDT to be completely sealed against the environment.

LVDTs are commonly used for position feedback in servomechanisms and for automated measurement in machine tools and many other industrial and scientific applications.

Ultrasonics

Ultrasonic sensors transmit sound pulses at a high frequency and evaluate the echo received back from the sensor. Sensors calculate the time interval between sending the signal and receiving the echo to determine the distance to an object.

Ultrasonic sensors are often used for distance measurement but are common in liquid and tank level applications. The technology is limited by the shapes of surfaces and the density or consistency of the material; for example, foam on the surface of a fluid in a tank could distort a reading.

Because of the effect of the air medium on the speed of sound on the signal, ultrasonic sensors are not particularly repeatable or precise; however, they can be used over fairly long distances and tend to have a smoothing or averaging effect when measuring irregular or moving surfaces.

Distance and Dimensions

Photoelectric sensors, proximity switches, LVDTs, ultrasonics, and encoders can all be used to measure distance and dimensions.

With optical sensors such as photo-eyes, the property of reflectivity can be used to determine the relative distance of an object from the sensor. As an object moves farther away, the amount of light received by the sensor becomes less. The color of the target also has an effect on the received signal, however, so optical distance measurement is best used on a consistent target. Laser-based devices can be used similarly to LED photoelectrics with longer range and less dependence on color.

Rows of LEDs or lasers that can measure dimensions based on the number of beams broken or the amount of light received and CCD-based devices that can measure distances accurately are other usable optical methods. These methods do not depend on reflectivity and can be used to measure nearly any object as long as it is not too large. Techniques using precision tooling and physical contact with the target such as LVDTs are also commonly used where contact with the target is feasible.

For longer-distance strokes LVDTs may not offer enough accuracy for an application. An excellent option for measuring distance is time-based magnetostrictive position sensing. Magnetostriction uses a ferromagnetic measuring element known as a waveguide, along with a movable position magnet. The magnet generates a direct-axis magnetic field within the waveguide. When a current or "interrogation pulse" is passed through the waveguide, a second magnetic field is created radially around the guide. The interaction between the two fields generates a strain pulse that travels at a constant speed from its point of generation at the magnet (the measuring point) to the end of the waveguide. A sensor detects the pulse and generates a highly accurate positional reading through the electronics of a high-speed counter.

Magnetostrictive sensors provide an absolute position reading that never needs recalibration or homing after a power loss. This can be a significant advantage over using LVDTs and encoders. The only limitation of this technology is that it cannot be used for short-distance dimensional measurements; the minimum range is about 25 mm. A well-known manufacturer and the first to develop products using this technology is MTS Systems, developer of Temposonics sensors.

Thermocouples and Temperature Sensing

There are a variety of devices that can be used to measure temperature. One of the most commonly used is the *thermocouple*. A thermocouple is a junction between two different metals that produces a voltage related to a temperature difference. They are inexpensive and interchangeable, have standard connectors, and can measure a wide range of temperatures. The main limitation when using a thermocouple is its accuracy; system errors of less than one kelvin (K) can be difficult to achieve.

Any circuit made of dissimilar metals will produce a temperature-related difference of voltage. Thermocouples for measurement of temperature are made of specific alloys, which in combination have a predictable and repeatable relationship between temperature and voltage. This relationship is not linear, however, and the voltage curve must be linearized in the input instrument. Temperature loop controllers contain linearization algorithms for the most common types of thermocouples. Selection of the thermocouple type can be made by setting dipswitches or software parameters.

Different alloys are used for different temperature ranges and to resist corrosion. Where the measurement point is far from the measuring instrument, the intermediate connection can be made by extension wires, which are less costly than the materials used to make the sensor itself. Thermocouples are standardized against a reference temperature of 0°C; instruments then use electronic methods of cold-junction compensation to adjust for varying temperature at the instrument terminals. Electronic instruments also compensate for the varying characteristics of the thermocouple within the linearization algorithm and help improve the precision and accuracy of measurements. An example of a thermocouple is shown in Fig. 3.14; the probe at the bottom is the sensing element inside a protective "well," while the container at the top is the head, which contains the termination points for the thermocouple wire.

FIGURE 3.14
Thermocouple.

Thermocouples are widely used in science and industry; a few applications would include temperature measurement for kilns and injection molding of plastics, measurement of exhaust temperature of gas turbines or diesel engines, ovens, and many other industrial processes.

The most common type of thermocouple in use is the K thermocouple (chromel-alumel). This covers temperature ranges from –200°C to 1350°C. It is inexpensive and available in a variety of styles. J thermocouples (iron-constantan) are less popular than K because of their lower usable temperature range of –40°C to 750°C. Other types of thermocouple include E, N, B, R, S, T, C, M, and chromel-gold/iron. A table for the different types of thermocouples is located in App. F.

One note on thermocouple polarity: there is a polarity labeled + and – for connection to input terminals. Counter to the common thought that the red wire is positive in many DC circuits, red is always the negative lead for thermocouples. Not every thermocouple pair has a red wire, but when using the American National Standards Institute (ANSI) color code, the red lead will always be negative.

Thermistors are a type of resistor with resistance proportional to its temperature. Thermistors are usually made of a ceramic or polymer material. They have a high precision over a limited temperature range, typically –90°C to 130°C.

Resistance temperature detectors (RTDs) also change resistance proportionally with temperature, but are made of pure metals. They are useful over a wider temperature range than thermistors but are less accurate. RTDs and thermistors may both be used with standard analog inputs and an excitation voltage because of their linearity, unlike thermocouples, which must use a special input to linearize the signal.

Infrared thermocouples or infrared temperature sensors are used as noncontact methods of sensing temperature. They use the thermal emission from the target to scale temperature to a readable value. They are usually manufactured to be used in place of a J- or K-type thermocouple for convenience.

3.3.3 Special-Purpose Sensors

There are various sensing devices that do not meet the criteria of being either digital or analog as they use elements of both.

Encoders and Resolvers

An *encoder* is a type of transducer that senses position or orientation, usually for use as a reference or active feedback to control position. Encoders may be rotary or linear, optical or magnetic, analog or digital, depending on the type of application.

Rotary optical encoders use a rotating glass or metal disk with slots or perforations along the circumference. An LED emits light along the path of the slots creating a train of pulses that can be used to count or measure distance. Figure 3.15 shows an open-style encoder.

Encoders may be open as in the illustration but are usually housed in a rugged metal housing with bearings and shaft for attaching to a motor with a coupling. The housing is generally watertight and may have either a potted cable or a multipin connector for termination.

By placing two sets of slots 90° out of phase with each other, the direction of rotation can also be determined. These two signals are known as the A and B pulses of the encoder. The inverse of the encoder A and B pulses are also often used, commonly known as A not and B not. A single slot is also placed along the circumference and is known as the Index or Z pulse; this is used for identifying the home or reference position of the encoder or device attached to it. This offset A and B pulse configuration is known as *quadrature*. This configuration is illustrated in Fig. 3.16.

Encoders are often of the multiturn variety; that is, they will turn multiple times, providing a count much higher than the number of slots on the disk. This means that the high-speed counter or servo module that the encoder is connected to must keep track of the number of turns or total count of the pulses. If the power is removed from the counter or control system, it is necessary to "home" the axis or device attached to the encoder, typically to an external "home" sensor and the index pulse.

Absolute encoders use a parallel signal to provide a binary count of the position of the encoder. The signal from an absolute encoder gives

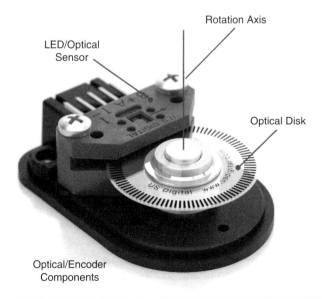

FIGURE 3.15 Encoder. (*Courtesy of U.S. Digital.*)

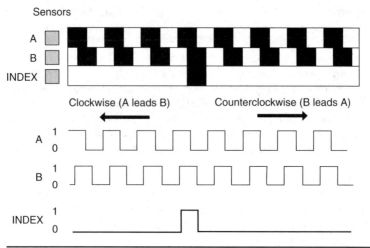

FIGURE 3.16 Incremental encoder disk quadrature track patterns.

an unambiguous position within its travel range without requiring knowledge of any previous position. This means that the encoder will have a fixed range or number of turns. Absolute encoders are often used when a system or axis must retain its position even when powered off and moved. Absolute and incremental encoders provide the same accuracy, but the absolute encoder is more robust to interruptions in transducer signal.

Resolvers are also used to detect rotary position and velocity. A resolver is best described as a rotating electrical transformer that provides a sinusoidal output, which is then converted to a digital value representing position. A common type is the brushless transmitter resolver. This type of resolver is similar to an electric motor in that it has a rotor and stator. The stator portion is made up of an exciter winding and two two-phase windings, labeled X and Y. These windings are located at 90° angles to each other. When an alternating current is induced into the exciter winding, the signal is transferred into the rotor windings and then back into the X and Y windings. This provides a sine and cosine feedback current, which is measured to determine the angle of the rotor. On one full revolution, the two feedback signals repeat their waveforms. Because resolvers are analog, they effectively have infinite resolution. Figure 3.17 shows a cutaway resolver in an industrial housing.

Vision Systems

Also known as machine vision, vision systems apply microprocessor or computer-based vision processing to inspection, measurement, and guidance tasks. While computer vision is mostly focused on image processing, machine vision may also require digital I/O

FIGURE 3.17   Resolver.

devices to control other manufacturing equipment. Machine vision is used in the inspection of manufactured goods such as semiconductor chips, automotive parts, food, and pharmaceuticals. It is also often used as a guidance method for robots.

Just as human inspectors working on assembly lines visually inspect parts to judge the quality of workmanship, so machine vision systems use smart cameras or digital cameras with computer-based image-processing software to perform similar inspections. Individual characteristics of parts can be assigned parameters to judge on a pass-fail basis for absence/presence, measurement tolerances, color, surface defects, and a number of other visually determined aspects.

Machine vision systems are also programmed to perform simpler tasks, such as counting objects on a conveyor, reading serial numbers, and measuring parts. Manufacturers favor machine vision systems for cases that require high-speed, high-magnification, 24-hour operation, and/or repeatability of measurements. Vision systems are more consistent than human beings because of distraction, illness, and other physical or mental limitations; humans are better at making finer qualitative judgments and adapting to new undefined defects.

Computers do not "see" in the same way human beings do. Cameras are not equivalent to human optics. Computing devices see by examining the individual pixels of images, processing them, and attempting to develop conclusions with the assistance of knowledge bases and features, such as pattern recognition engines. Although some machine vision algorithms have been developed to mimic human visual perception, no machine vision system can yet match the capabilities of human vision in terms of image comprehension, tolerance to lighting variations, image degradation, and part variability.

A number of unique processing methods have been developed to process images and identify relevant image features in an effective and consistent manner. Among these are various line, circular, or area tools to detect edges or count pixels within a defined intensity or brightness range; "blob" tools to identify patterns or shapes of a certain size; and color and text recognition tools.

A typical machine vision system will consist of several of the following components:

1. A digital or analog camera (black-and-white or color) with optics for acquiring images.

2. Camera interface for digitizing images (widely known as a "frame grabber"). This converts the image to a digitized format, typically a two-dimensional array of intensity values. This is then placed in memory for analysis by the software algorithms.

3. A processor (often a PC or embedded processor, such as a digital signal processor [DSP]).

4. I/O hardware (digital I/O) or communication links (usually a network or RS-232 connection) to report results.

5. A lens to focus the desired field of view onto the image sensor.

6. Suitable, often very specialized, light sources (LED illuminators, fluorescent or halogen lamps, direct on-axis, and others). The lighting is designed to enhance or highlight certain features while obscuring or minimizing those that are not of interest. Generating or eliminating shadows is one of the principal purposes of adding lighting.

7. A program to process images and detect relevant features.

8. A synchronizing sensor for part detection (often a photo-eye or proximity switch) to trigger image acquisition and processing. This sensor may also be used to trigger a synchronized lighting pulse to freeze a sharp image.

In some cases, some or all of the above are combined within a single device, called a smart camera. The use of an embedded processor eliminates the need for a frame grabber card and external computer, reducing cost and complexity of the system while providing dedicated processing power to each camera. Smart cameras are typically less expensive than systems made up of a camera and a board and/or external computer. Figure 3.18 shows two different types of camera.

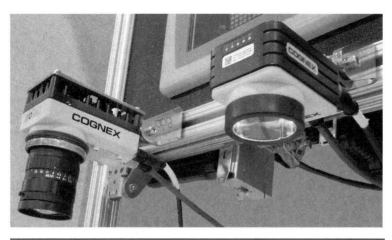

FIGURE 3.18 Cognex In-Sight smart cameras.

The one on the right is used for reading bar code images. Note the adjustable lens on the left camera; this is used to regulate the light input and focus the image.

The camera itself typically uses a CCD or CMOS image sensor. Both of these devices perform the task of converting light into an electrical signal. The array of pixels creates an image by patterning the light and dark pattern focused onto the sensor by the optics of the camera. The intensity levels are then processed by the software into patterns that can be analyzed by the various tools.

Usually the software takes several steps to process an image. First the image is processed to reduce noise. It may also convert the many analog shades of gray into a simpler combination of black-and-white pixels, a process known as binarization. To do this an analog threshold is set in the software. After this simplification of the image, software can count or identify objects and measure or determine the size of patterns and features. The final step is to pass or fail the captured image based on the criteria entered by the user. The result is then communicated by digital signals or communications to a control system that can then act on the information to reject or process the part.

Though most machine vision systems rely on black-and-white cameras, the use of color cameras is becoming more common. It is also increasingly common for machine vision systems to include digital camera equipment for direct connection rather than a camera and separate frame grabber, thus reducing signal degradation.

X-ray sensors are sometimes used to look inside materials for flaws such as cracks or bubbles. When combined with vision technology, these sensors can be used for automatic material sorting.

Gas Chromatography

Gas chromatography-mass spectrometry (GC-MS) is a method that is used in some chemical and process plants as a means of identifying and separating substances. This requires that a sample of the substance be captured, ionized, accelerated, deflected, and detected at the molecular level. The instruments that do this are quite expensive but are used in the food and beverage, perfume, and pharmaceutical industries.

Bar Codes, RFID, and Inductive ID

A *bar code* is a method of representing data by putting it into a visible, machine-readable format. Originally, bar codes were only represented by parallel lines that varied in width and spacing to encode alphanumeric data. This is called one-dimensional (1-D) or linear bar coding. Two-dimensional methods are also widely used today as the reader technology has evolved.

Linear or 1-D readers contain a light source that reflects off the black-and-white lines similar to a diffuse photo-eye. The light source is generally a red LED or laser. To cover a larger read area, the transmitted light will sometimes "raster" or move up and down. Figure 3.19 shows a commonly seen linear bar code.

The mapping of patterns into characters is known as a "symbology." This specification includes the coding for the alphanumeric characters along with the start and stop characters and computation of a checksum (a simple error-detection scheme).

There are more than 30 different 1-D codes in use. Most of these fall into two groups, discrete or continuous, depending on whether characters begin and end with a bar or not. There are also two-width or many-width classifications. Some of the more common symbologies are UPC, Code 39, and interleaved 2 of 5. Most 1-D readers can be set to read any of the common formats.

Figure 3.19 1-D bar code.

Bar codes later evolved into other geometric patterns in two dimensions (2-D). These bar codes are usually made up of rectangles, dots, hexagons, or other geometric shapes arranged in a grid pattern. Readers for 2-D bar codes generally use a CCD camera to capture the bar code image. Two-dimensional symbologies cannot be read by a laser as there is typically no sweep pattern that can encompass the entire symbol. Figure 3.20 shows a 2-D bar code.

Some of the more common codes for 2-D symbologies are DataMatrix, Codablock, EZCode, and QR code. The automotive industry is a major user of 2-D DataMatrix codes since the pattern can be directly imprinted into a metal part using a pinstamp or "dot peen" marking system. Laser etching can be used for the same purpose.

Radio frequency identification (RFID) systems are another method of tagging parts and identifying them. Unlike bar codes, however, the tag does not have to be within line of sight range of the reader and may even be embedded inside an object. A common use of RFID systems in industrial automation is to track pallets or carriers through a process. An RFID system consists of a radio transmitter-receiver for two-way communication interfaced with a processor for the received information and the RFID tags that contain the information. Tags consist of an integrated circuit containing data and an antenna. The RFID reader sends a signal to the tag and reads its response. It may also work as a read-write system that transmits data to the tag for tracking purposes.

RFID tags can be passive, using the radio energy transmitted by the reader to power its circuit, or active with a tiny battery. Another option is a battery-backed passive tag, which is only activated when in the presence of a reader. Passive tags can be made much smaller and less expensively than active or battery-backed passive tags but must be very close to the reader for the field to be strong enough to activate the tag. Figure 3.21 shows a reader and several RFID tags.

FIGURE 3.20
2-D bar code.

92 Chapter Three

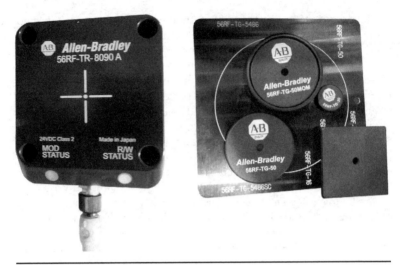

FIGURE 3.21 RFID reader and tags.

Tags may contain a precoded unique serial number for lookup in a database or hold product-related information such as a part number, production date, or lot code. Read-write tags may be coded at various locations as a part travels through production, or they may be of the write-once read-multiple variety, also known as a field programmable tag. Information coded into a tag is stored electronically using nonvolatile memory.

RFID systems usually operate either in the high-frequency (HF) or ultrahigh-frequency (UHF) range of the radio spectrum. The distance at which an RFID tag can be read varies from less than a foot for some inexpensive, small passive tags to hundreds of feet for some larger active tags. More than one tag at a time may respond to the interrogation signal transmitted by a reader, so collision detection is often an important feature for an RFID controller.

Inductive ID systems serve a similar function to RFID systems but use a coil of wire similar to a proximity switch. The reader will excite an oscillator circuit in the tag that will transmit a serial code. Inductive ID systems can be lower cost and less susceptible to radio interference, but typically handle less information. They also have a much shorter range. Like RFID tags, inductive tags also come in active or passive, read-write, or read-only varieties.

Keyboard Wedge

A keyboard wedge is an interface that allows a device such as a bar code scanner or magnetic strip reader to emulate a keyboard. The name "wedge" describes the physical position it occupies wedged between the keyboard and the computer port—Fig. 3.22 illustrates this arrangement. For example, a bar code reader converts the scanned

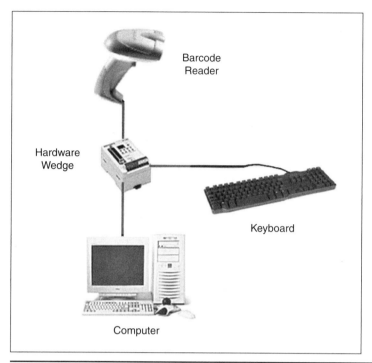

FIGURE 3.22 Keyboard wedge diagram.

code into a human readable alphanumeric format and then passes it through the wedge as if it were typed on the keyboard. The computer does not know whether the data came from the keyboard or another device, and the data is translated seamlessly.

A keyboard wedge may also be a software program that takes information into a USB or COM port and routes it through the keyboard buffer. Again, this is a transparent process from the perspective of the computer. This is a less expensive method of interfacing card scanners or bar code readers with a control system, but typically designers would choose a dedicated port for peripheral devices that are to be used often.

3.4 Power Control, Distribution, and Discrete Controls

From the service entrance of most industrial facilities power is distributed by way of three-phase busway or wired into distribution panels. Usually voltage at the service entrance is reduced via transformer to three-phase 480VAC. Various fused disconnects or circuit breakers are located to provide protection for branches of the distribution system. Disconnects that can only be reached by long poles with hooks at the end are often located at the point of power drops to individual machines. Wiring is distributed inside rigid conduit or EMT to fixed locations or in flexible Seal-Tite or power cable to more temporary or movable spots. Cable tray is often mounted from the ceiling with multiconductor control or power cable laid in it and dropped to points of use. It is important to consult the National Electric Code or local regulations when planning a power distribution system.

Wire and cable are sized by the allowable amount of current that they are rated to carry for certain conditions, such as temperature or insulation. Wire is rated by gauge and may be sized in terms of American wire gauge (AWG), standard wire gauge (SWG), or imperial wire gauge. Wire is usually rated by Ampacity, another name for the amount of current it can safely carry.

Relays and contactors are a form of switching device that applies or removes power from a circuit based on a remote or external signal. Timers and counters also switch power or signals based on a delay or set number of pulses.

3.4.1 Disconnects, Circuit Breakers, and Fusing

An individual line of automation equipment or single machine will typically have one main disconnect to allow power to be removed from a single source. These disconnects usually have a set of fuses or a circuit breaker rated appropriately for the equipment they are supplying. There will often be several levels of branch circuit protection after the main disconnect in the form of fuses or circuit breakers. In most cases these also serve as a manual disconnect for the branch, although some motor circuits will simply have a fuse clip with no disconnecting means upstream. There are regulations concerning disconnects being present within a certain distance from a motor, so disconnects without fusing are sometimes located nearby for quick power removal. Disconnects consist of a set of contacts rated for the amount of current they must break with a manual means of actuation. These may also include a means of remote actuation or control.

Circuit Breakers

A circuit breaker is a circuit protection device that can be reset after detection of an electrical fault. Like all circuit protection devices, its purpose is to remove power from an electrical device or group of devices, protecting the circuit from damage. Circuit breakers are rated by the current at which they are designed to trip, as well as the maximum current they can safely interrupt during a short circuit.

Circuit breakers interrupt a current automatically; this requires some kind of stored mechanical energy, such as a spring or an internal power source, to actuate a trip mechanism. Small breakers such as those used for branch circuit or component protection in a machine are usually self-contained inside a molded plastic case. Larger circuit breakers usually have a pilot device that senses a spike in current and operates a separate trip mechanism. Current is detected in several ways. Magnetic breakers route the current through an electromagnetic circuit. As the current increases, the pulling force on a latch also increases, eventually letting the contacts open by spring action. Thermal magnetic circuit breakers use a bimetallic strip to detect longer-term over current conditions while using a magnetic circuit to respond instantly to large surges, such as a short circuit.

Circuit breakers usually have a reset lever to manually trip and reset the circuit. This is an advantage over using fuses, which must be replaced after one use. In industrial applications most circuit breakers are used for low-voltage application (under 1000 V). Medium-voltage (1000 to 72 k) and high-voltage (more than 72.5 kV) breakers are used in switchgear applications but are rarely seen in industrial plants, though medium voltage switchgear is used in some process facilities. Low-voltage breakers may be of the DC or AC variety and generally fall into the categories of miniature circuit breakers (usually DIN rail mounted, up to 100 A) and molded case circuit breakers (self-contained, up to 2500 A). An example of a molded case breaker is shown in Fig. 3.23.

Circuit breakers must carry the designed current load without overheating. They must also be able to withstand the arc that is generated when the electrical contacts are opened. Contacts are usually made of copper or a variety of alloys. Contact erosion occurs

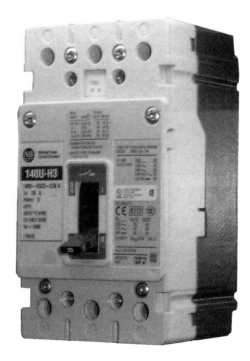

FIGURE 3.23
Molded case circuit breaker.

every time the contacts are opened under load; usually miniature circuit breakers are discarded when the contacts are worn, but some larger breakers have replaceable contacts.

There are two types of trip units in low-voltage circuit breakers: thermal magnetic and electronic. The thermal magnetic trip units contain a bimetallic thermal device that actuates the opening of the breaker with a delay depending on the overcurrent value. These are used for overload protection. The magnetic trip device has either a fixed or adjustable threshold that actuates the instantaneous trip of the breaker on a predetermined overcurrent value—usually a multiple of the overload current rating.

Electronic trip units use a microprocessor to process the current signal. Digital processing provides four different trip functions: the long and short time-delay trip functions per ANSI code 51 (AC time overcurrent), the instantaneous trip function per ANSI code 50 (instantaneous overcurrent), and the ground-fault trip function per ANSI code 51 N (AC time ground fault overcurrent).

Circuit breakers are categorized by their characteristic curves for different applications. Highly inductive loads, such as transformers, can have very high inrush currents of 10 to 20 times the current rating of the device. These are classified as a class "D" curve. Normal inductive loads, including most motors, have a current inrush rating of 5 to 10 times the rating of the device and are classified as a class "C" curve. A class "B" curve is used for most lighter-duty noninductive loads and has a rating of two to five times the circuit breaker rating.

Circuit breakers are also rated for use as branch, supplementary, or feeder devices. Feeder circuit breakers are generally of the molded-case variety and are designed for main power feeds. They are typically tested at 20,000 A interrupting rating. Branch circuit protection is tested for at least 5000 A interrupting rating and is used for branch circuits under the main breaker. Feeder and branch circuit breakers must be listed devices by Underwriters Laboratories (UL).

Supplementary protection devices are used for equipment protection in a branch circuit. They are classified as "recognized components" by UL rather than listed. They are tested with upstream branch circuit protection and are generally rated for 5000 A or less.

Motor circuit protectors (MCPs) are special application breakers with adjustable magnetic settings. They allow the operator to set the breaker's magnetic protection level just above the inrush level of the motor. Overload protection for the motor is supplied in the starter's overload relay. This combination allows protection of the motor without causing nuisance trips. MCPs are UL-recognized components.

Motor protector circuit breakers (MPCBs) are UL-listed circuit breakers with fixed magnetic protection and built-in motor overload protection. These breakers' trip units are adjustable for Motor FLA ratings and can be set for overload trip class. MPCBs can be used directly with a contactor for a complete motor starting and protection package.

Fuses

A fuse or fusible link is an overcurrent protection device that is designed to melt (or "blow") when excessive current flows through it. It is composed of a metal strip or wire element rated at a specified current plus a small percentage. This is mounted between two electrical terminals and generally surrounded with a nonflammable insulating housing.

Fuses are placed in series with the current flow to a branch or device. If the current flow through the element becomes too high, enough heat is generated to melt the element itself or a solder joint within the fuse.

Dual element fuses contain a metal strip that melts instantly for a short circuit as well as a low melting solder joint for longer-term overloads. Time delay or "slow blow" fuses allow short periods of overcurrent conditions and are used for motor circuits, which can have a higher current inrush as the motor starts.

Fuses are made in many different shapes, sizes, and materials, depending on the manufacturer and application. While the terminals and fuse element must be made of a metal or alloy for conductivity, the fuse body may be glass, fiberglass, ceramic, or insulating compressed fibers. Fuse sizes and mounting methods generally fall into several standardized formats. Figure 3.24 shows several cartridge-type fuses; note that the larger fuse on the left has an indented area or

FIGURE 3.24   Cartridge fuses.

groove at the bottom end of the fuse. This is known as a rejection fuse, the feature ensures that the fuse can only be placed into its holder one way.

Most fuses used in industrial applications are cartridge fuses. These are cylindrical with conducting caps on each end separated by the fusible link covered by the housing. These may be small glass or ceramic fuses for light loads or larger J or R class fuses. Cartridge fuses are also known as ferrule fuses. The caps may have bladed ends for insertion into clips or have a hole in the blade for bolting to a terminal. The most common methods are spring clips or terminal block–style fuse holders.

Fuses for use on printed circuit boards (PCBs) are generally soldered into place. They may have wire leads or solder pads depending on the desired mounting technique.

Fuses Compared with Circuit Breakers

Fuses are less costly than circuit breakers but must be replaced every time an overcurrent event occurs. This is not as convenient as simply resetting a breaker, though, and makes it more difficult to ignore intermittent faults.

Fuses react more quickly than circuit breakers, especially the "current-limiting" variety. This helps minimize the damage to downstream equipment.

3.4.2 Distribution and Terminal Blocks

Cable and wire is distributed to multiple circuits by terminating the ends into a securing means such as a screw or clamp. *Distribution blocks* are used for larger gauges of wire. They are usually connected by means of screwing a threaded stud into a block of metal. One side will have one termination point and the other will have multiples to feed branch circuits. Distribution blocks are mounted into insulating carriers with dividers between phases and are available in one to four pole configurations. They can have one or more terminations for each pole on the incoming side and up to 12 terminations each on the outgoing or branch side. Figure 3.25 shows a three-phase open-style distribution block; there are six connections for each phase. Wire gauges start at around 14 AWG and go all the way up into large MCM-size cable. MCM is an abbreviation for thousand circular mil (a mil is a thousandth of an inch).

Multiterminal copper or aluminum bus bars with a series of screw terminals are often used for ground or neutral terminations in panels. These are typically for smaller wire sizes that have no voltage present.

Terminal blocks are used to make wiring and cable connections and manage wiring. They are sized for the ranges of the wire and cables that are to be connected. Screw terminal and spring clamp types are both widely used on smaller conductors, but for large wire sizes screw terminals are usually used.

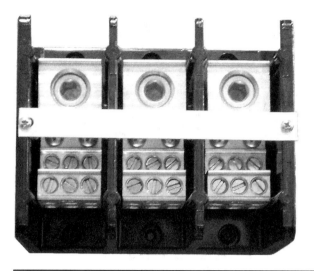

FIGURE 3.25 Distribution block.

There are a wide variety of styles and manufacturers of terminal blocks. The main purpose of terminal blocks other than wire management is to insulate the exposed wiring ends when making connections. The National Electrical Manufacturers Association (NEMA) has a number of standards associated with terminal block specifications as does the International Electrotechnical Commission (IEC). Terminal blocks are often categorized as being NEMA or IEC style. NEMA terminal blocks are typically a more open style, while IEC is considered a "finger-safe" style with insulation surrounding the screw or clamp terminals.

Terminal blocks are generally sized to mount on uniform-size metal rails known as DIN rail. DIN is an abbreviation for Deutsches Institut für Normung, a German standard. Figure 3.26 shows a selection of various types of labeled IEC terminal blocks mounted on a piece of DIN rail. The larger black block is a cartridge fuse-holding terminal block.

Usually connections are straight through the block from terminal to terminal, but removable jumpers, switches, or fuses are sometimes built into the block. Terminal blocks for fuses are also called fuse blocks. These are made to swing open for fuse removal and also double as a branch or component disconnect. LED indicators are also embedded in some terminal blocks for energy presence or blown fuse indication. Special-purpose terminal blocks with contacts for thermocouples and extremely low or high voltages are also available.

Terminal blocks are commonly available in one-, two-, and three-level configurations as a space-saving feature. They are usually mounted to some type of metal rail, the most common being DIN rail.

100 Chapter Three

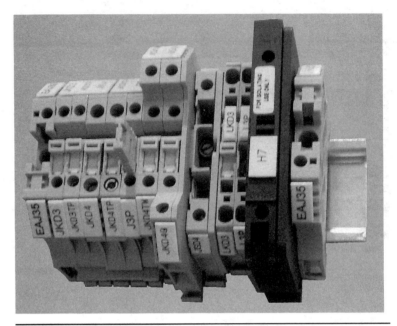

FIGURE 3.26 IEC Terminal blocks on DIN rail.

This allows blocks from different manufacturers to be mounted on a common surface. They are made in a wide variety of colors, which are often used for circuit identification.

A wide range of accessories are available for terminal blocks, including center and side jumpers to form a common bus, DIN rail, labels and labeling kits, and end caps and anchors.

3.4.3 Transformers and Power Supplies

Transformers are used to isolate or transfer energy in the form of AC current from one circuit to another. This is done by the principle of mutual inductance. If a changing current is passed through a coil of wire, it creates a magnetic field that can be used to create a current in another coil of wire that is electrically isolated from the first coil. This is most often accomplished by wrapping both coils around a common core of iron rich metal.

One of the principles of this induced voltage is that the voltage can be raised or lowered in proportion to the number of turns in the coils. A formula that can be used to express this relationship is Vp/Vs = Np/Ns, where V is voltage, N is the number of turns in the coil, p is the primary or the coil where the voltage is applied, and s is the secondary where the converted voltage is applied to the load. A transformer that is used to increase the voltage from the primary to the secondary is known as a "step-up" transformer, and the opposite

Components and Hardware   101

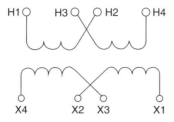

Connections		
Primary Volts	Interconnect	Primary Lines Connect To
480 V	H2-H3	H1, H4
240V	H1-H3, H2-H4	H1, H4
Sec. Volts	Interconnect	Secondary Lines Connect To
120 V	X1-X3, X2-X4	X1, X4
240 V	X2-X3	X1, X4

FIGURE 3.27 Transformer wiring diagram.

is a "step-down" transformer. Of course, by Ohm's law, when the voltage is increased, the current will be decreased accordingly and vice versa.

Figure 3.27 is a wiring diagram for a single-phase power transformer. Transformers can typically be "tapped" or wired in different ways, as shown in this diagram. This particular transformer can provide 480 to 120, 480 to 240, 240 to 120, and 240 to 240 voltage conversions.

Transformers are also used for isolation purposes, as shown in the 240 to 240 wiring (which could also be 120 to 120). Since a voltage cannot change instantaneously through an inductor, isolation transformers are often used to protect the load from quick spikes in a circuit. They are commonly used in control and drive systems.

Transformers come in a wide range of sizes, from small internally mounted transformers inside devices such as DC power supplies to large three-phase transformers that power an entire production line or section of a plant. Many commercially available transformers have multiple taps, allowing the same transformer to provide a range of voltages depending on how these taps are connected.

Another method of obtaining different voltages from one transformer is to use an *autotransformer*. This is a transformer that only has one winding with taps on each end and one at an intermediate point. Voltage is applied to two of the terminals. The secondary is then taken from one of the primary terminals and the third terminal. The location of the intermediate tap determines the windings ratio and therefore the output voltage. If insulation is removed from part of the windings, the intermediate tap can be made movable using a sliding brush, making the output voltage variable similar to a potentiometer.

The purpose of a transformer in an automation system is either to convert an AC voltage to a different voltage for distribution in the system or to isolate a circuit from another. Commonly three-phase 480VAC is applied to the disconnect of a control enclosure. From the disconnect power is distributed through various branches with circuit protection for different purposes. Where a lower voltage (usually 240,208 or 120VAC) is required, a transformer is connected to reduce the voltage level. Transformers may be used between individual phases to develop a single-phase voltage or across all three phases. Windings are often tapped in the center and grounded to develop two phases 180° out of phase with each other. This is similar to a common residential service entrance, where 240VAC is wired to a distribution or breaker panel along with a neutral (the grounded center tap). The 240 V can then be used for higher-power appliances and two rows of breakers supply 120VAC to branch circuits. Transformers typically have circuit protection such as fuses or power supplies on both the primary and secondary side.

DC *power supplies* are used to provide lower-voltage DC power for I/O devices such as sensors and solenoid valves. Power supplies usually have regulated outputs to prevent current or voltage fluctuations. They are usually protected on the AC and DC sides by fuses or circuit breakers.

The most common voltage used for industrial machinery is 24VDC. This is a low enough voltage to prevent most injuries but high enough to minimize noise interference and allow distribution over a reasonable distance. 12VDC is also sometimes used, while higher DC levels of 48 or more volts may be used for DC motors such as steppers. Servos and DC motors do not usually use separate power supplies, but generate their own DC power in the drive.

3.4.4 Relays, Contactors, and Starters

A relay is a device that allows switching of a circuit by electrical means. There are various types of relays, including electromechanical and solid-state coils, reed or mercury wetted contact, but the purpose is generally the same: to control a circuit with one voltage with a signal from another or to use one signal to switch multiple circuits, as in Fig. 3.28.

Electromechanical relays use an electromagnetic coil to physically pull a set (or sets) of contacts either from an open to a closed position or from closed to open. AC or DC may be used to switch the coil; this is one of the specifications of a relay along with the number of poles and amount of current that can be handled by the contacts. Contacts are specified as NO or NC, referring to their deenergized state. Relays may have multiple poles of both NO and NC contacts. Figure 3.29 shows a variety of different relay types; the relay to the far left is a

Components and Hardware **103**

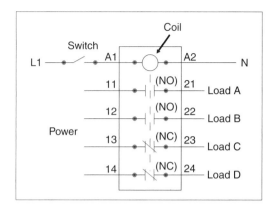

FIGURE 3.28
Schematic of four pole relay.

FIGURE 3.29   Relays.

tube-base electromechanical relay and socket, and the next two are often referred to as "ice-cube" relays. The relay on the lower right is a heavier-duty DIN rail mount electromechanical relay, while the one at the upper right is an adjustable timing relay.

Solid-state relays use transistor technology to switch current flow. Voltage is applied to a solid-state "coil" that may switch current directly through a transistor or CMOS device or energize an LED to optically isolate the circuits. Solid-state relays have no moving parts, which gives them greater longevity than electromechanical relays; however, they are rated at a lower current-switching capability.

Some relays have a coil to latch the relay on and a separate coil to reset it. These are used when a circuit state needs to be maintained even if power is lost. These are known as latching or set-reset (SR) relays.

Safety circuits often use relays that have force-guided contacts. This means that the contacts are mechanically linked together so that all of them switch together. This ensures that if a set of contacts weld together because of arcing, then one set of contacts can be used to reliably monitor the state of the relay. These safety relays also use redundant sets of contacts for each circuit for the same reason.

Relays come in a variety of form factors also. Large relays generally are screw or bolt mounted directly to a panel or backplane, while many standard industrial relays have round pins or blades that can be plugged into a DIN rail–mounted socket. Sockets are available in tube base for round pins and bladed base or pin sockets for small relays. Small relays may also be soldered to a circuit board.

A type of relay that can handle the high power required to directly control an electric motor is called a *contactor*. Continuous current ratings for common contactors range from 10 A to several hundred amps. Contactors are an element of motor starters; a *motor starter* is simply a contactor with overload protection devices attached. The overload sensing devices are a form of heat operated relay where a coil heats a bimetal strip, or where a solder pot melts, releasing a spring to operate an auxiliary set of contacts. These auxiliary contacts are in series with the coil. If the overload senses excess current in the load, the coil is deenergized.

Motor starters are generally categorized as NEMA or IEC style. NEMA starters are generally larger and have replaceable overload elements. They can generally be rebuilt if necessary; however, they are physically larger and more expensive than an IEC motor starter of the same rating. IEC starters are not usually rebuilt and are simply discarded when the contacts wear out. Figure 3.30 is an IEC motor starter in a manual motor control enclosure.

3.4.5 Timers and Counters

A timer reacts to an applied signal or power feed and switches a set of contacts based on a delay. It may also create a repetitive series of pulses. Timers may be purely mechanical, such as with a pneumatic timer; electromechanical with a motor and clutch; or entirely electronic. They are available in both analog and digital formats.

Timers generally fall into the following categories:

- On Delay—Timer changes state after a specified period of time and remains in that state until the signal is removed.
- Off Delay—Timer changes state immediately and reverts to its original state after a specified period of time.
- One Shot—Timer creates a single pulse of specified length.

Components and Hardware 105

FIGURE 3.30 IEC motor starter.

- Pulse or Repeat-Cycle—Timer creates a series of on and off pulses with configurable on and off times until signal is removed.

As with temperature controllers, timers and counters are often sized using the DIN system, ensuring that they will fit a certain-size panel cutouts. They are usually available in 1/16 DIN, 1/8 DIN, or 1/4 DIN sizes. Figure 3.31 shows a 1/16 DIN digital timer.

Electromechanical timers such as the Eagle Signal Cycle Flex timer shown in Fig. 3.32 are often used in applications where electronic timers may not be appropriate. Mechanically switched contacts may still be less expensive than the semiconductor devices needed to control powerful lights, motors, and heaters. An electromechanical cam timer uses a small synchronous AC motor turning a cam against a bank of switch contacts. The AC motor is turned at an accurate rate by the applied frequency, which is regulated very accurately by the power companies. Gears drive a shaft at the desired rate and turn the cam. These timers are still in use in many industrial facilities because

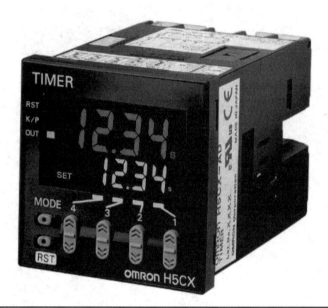

Figure 3.31 1/16 DIN digital timer. (*Courtesy of Omron.*)

Figure 3.32
Eagle Signal electromechanical timer.

they are easily rebuilt, rugged, and switch high-current loads; however, they are often replaced with less expensive and more reliable electronic timers.

The most common application of electromechanical timers now is in washers, driers, and dishwashers. This type of timer often has a friction clutch between the gear train and the cam, so that the cam can be turned to reset the time. This is a less expensive method of performing multiple timing segments with high-current load switching than with an electronic version.

A counter also reacts to input signals, totalizing them and changing the state of a signal when the specified count has been reached. Counters are generally classified as up counters, where state changes increment the value until the set point is reached, or down counters, where the counter starts at the set point and counts down to 0. Counters may also be combinational with both up and down signals. Counters also have a reset input to set the counter back to its starting point. They may be mechanical in nature, such as with a totalizer, or be incremented electronically.

3.4.6 Push Buttons, Pilot Lights, and Discrete Controls

Before the advent of touch screens, signaling and machine control had to be done with push buttons, switches, and indicator lights. A large variety of components are still used as discrete interfaces between an operator and a machine today.

Push Buttons and Switches

A *push button* is a manually operated spring-loaded method of opening or closing a set of electrical contacts. Industrial push buttons generally come in several standard sizes; 30 mm, 22 mm, and 16 mm diameters. There are larger and smaller sizes also available, but the vast majority of push buttons, switches, and pilot lights fall within these standard sizes.

Larger push buttons (22 and 30 mm) are often modular in nature, having an actuator to which contact blocks can be mounted and removable mounting rings and bezels. Contact blocks are available in NO and NC configurations and may be mixed and matched as necessary. These blocks can also be stacked on top of each other for up to four sets of contacts. Push buttons may also have an internal light that may be illuminated from a control output or through one of the sets of contacts; these are usually LED or incandescent bayonet base bulbs.

Push buttons come in various colors, generally black, red, yellow, green, blue, or white, although other colors are also sometimes seen. The actuator may be mushroom head, extended, or flush with the bezel. They also may be of the momentary (spring return) or maintained (toggling) variety. Figure 3.33 is a 30 mm flush push button.

Selector switches have many of the same characteristics as push buttons; they use the same types of contact blocks and come in the same 16, 22, and 30 mm diameters. The color is usually black, although inserts may be of various colors. Lights are not commonly used in selector switches.

As with push buttons, switches may be of the maintained or spring return variety. Most switches have two or three positions, although four positions or even more are sometimes seen. Unlike push buttons, however, all of the contacts do not switch at the same time. For a three-position switch the contacts on one side will switch

FIGURE 3.33 30 mm push button.

in the left position and the opposite side will switch in the right position. Both sides usually remain unswitched in the center. Contact blocks are actuated with a cam rotating with the switch body.

Part of the push button or selector switch assembly is usually an antirotation ring. This is a ring with a tab that fits into a slot in the device and also a tab that fits into a slot cut into the punched hole in the cabinet or enclosure.

Pilot Lights and Stack Lights

Pilot lights are available in the same standard sizes as push buttons and switches, although much larger lights are sometimes used for greater visibility and small 8 mm and 10 mm lights are common for higher-density display. Lamps for pilot lights are generally of the incandescent or LED type. Most lamps are white and a plastic cover is used to change the color of the light. They are available for "full voltage" applications of 120 to 240VAC, which use a small transformer, as well as 12 and 24VDC. For low-voltage or computer card outputs, 5 and 6VDC lamps are often used. Some pilot lights also have a spring-loaded "push to test" feature that will illuminate the lamp, although there are no external contacts for these as there would be on a lighted push button. If pilot lights are connected to controlled outputs, a separate "push to test" push button is sometimes used to illuminate all of the lamps on a panel at once.

Stack lights, also called light stacks or tower lights, are columnar sets of lights that usually indicate the state of a whole machine or control system. They are also modular, usually beginning with a base unit that may or may not include a horn or buzzer. The base may be connected using a quick disconnect cable or terminals with a strain relief entry. Light units are then stacked onto the base in the required order, generally up to five units in height. A common combination of lights would be red, yellow, and green (from top to bottom), red generally signifying a fault or alarm, yellow signifying either caution or manual/maintenance mode, and green signifying auto mode or machine running. There is no universal stand for these colors, and each company or plant may have their own specifications. Blinking the lights to signify auto/not started or cycle stop versus immediate stop conditions is an example of how a stack light might be used to deliver additional information. Blue or white lights are sometimes added for signals like low bin or hopper or other special functions as defined by the designer. Figure 3.34 shows several arrangements of four-color stack lights; the two rightmost stacks have a buzzer or audible alert in the top position.

FIGURE 3.34 Stack lights. (*Courtesy of Banner.*)

As with pilot lights, they may be energized by 24VDC, 120VAC, or various other voltages as required. Other modules for playing a recorded voice or music are also available. A feature of some stack lights that may be beneficial is a flexible base to reduce the chance of the stack breaking off a low-mounted cabinet or machine. Stack lights may also be pole mounted and side mounted, depending on the application.

Other Panel-Mounted Devices
Some items grouped under the heading "discrete controls" may not be discrete at all. An example is a panel-mounted potentiometer for analog speed control of a motor drive. These are often available in the same form factors as selector switches in 22 mm or 30 mm sizes. Temperature controllers, timers, and counters are also devices you might find mounted on the front of a controls enclosure.

Horns and buzzers are other discrete devices that may be panel mounted. Buzzers are generally piezoelectric and susceptible to moisture since they cannot be easily sealed.

Along with all of the devices that might be mounted on a controls enclosure come labels for these devices. Most commonly engraved plastic tags or painted metal tags that have the appropriate-size hole are used with push buttons, switches, and pilot lights. Engraved plastic or metal tags are of two colors, an inside and outside color. An example is Gravoply plastic, which may be black or red on the outside with a white inside color. When characters are cut into the plastic, the inside color shows through. Tags may be premade with common terms like *stop*, *start*, and so on, or be sold as blanks for the user to engrave.

Tags are not only used for devices on the outside an enclosure; they are also common inside the cabinet or mounted next to a sensor on a machine. These may be engraved or printed and contain schematic, I/O, or descriptive text for components. Safety warnings also fall into the premade or purchased label category.

3.4.7 Cabling and Wiring

An important part of the distribution of power and signals throughout a system is cable and wire. Individual wires and multiconductor cables are used to connect the various control devices and distribution components within a machine or system. Wire sizes or gauges are specified as described in the appendixes of this book. Wire may be made of any conductive metal but is usually copper or aluminum. Usually it is covered with a thermoplastic insulation available in a wide range of colors. Wire is manufactured in solid or stranded forms, depending on the application.

Multiconductor cables consist of a collection of insulated wires inside a protective jacket. The wires may be twisted together in pairs for noise immunity or simply run in parallel. Multiconductor cables

FIGURE 3.35 Multiconductor cables.

often carry a noninsulated *shield* or *drain* wire to help carry away unwanted stray signals. This wire should be grounded at *one end only* to prevent a ground loop. An additional foil covering is often wrapped around the bundle inside the jacket but in contact with the shield wire. Examples of several multiconductor cables are shown in Fig. 3.35.

For high-flexing and repetitive movement applications, multiconductor cable is often made with fine stranding to improve its bend radius and increase usable lifetime. Specifications for expected number of cycles and minimum bend radius are often listed in wiring catalogs.

Connecting individual wires or multiconductor cables together may be done with terminal blocks, but in some cases they must be spliced. This may be done using crimp on malleable metal pieces called butt splices or wires may be soldered. After soldering, the wire junction must be insulated using electrical tape or shrinkable tubing, also known as *heat shrink*.

Strain Relief

To prevent pulling wire and cabling out of terminations a strain relief–type fitting is placed at enclosure entry points and built into cable plugs. These may be of a screw clamp type or a rubber "donut" shape that clamps down on the cable when a fitting is tightened. Another type of strain relief has a series of ridges at the point where the cable meets the enclosure or junction box. The main purpose of a strain relief is to reduce wear or stress at the point of entry when a cable is pulled. Strain reliefs often also provide ingress protection from liquids. Strain reliefs often come in standard hole sizes like electrical fittings; 3/8 in, 1/2 in, and 3/4 in. One inch and larger are all standard sizes. They may be made of a galvanized metal or plastic. Figure 3.36 illustrates a 1/2-in cord grip strain relief installed in the side of an enclosure.

Ferrules

A ferrule is a circular clamp or sleeve used to hold together and attach fibers or wires by crimping the ferrule to permanently tighten it onto the wire end. Wiring ferrules often have a color-coded piece of plastic molded around one end both to allow easy wire entry and for identification of wire gauge. Ferrules prevent smaller stranded wires from splaying and provide a solid electrical connection for terminal block clamps or screws. Special crimping tools with selectable dies are used to crimp the ferrule firmly onto the exposed wire end. Figure 3.37 shows several different sizes of insulated ferrules.

Soldering

A common method of attaching wires to each other or to pins in plugs is soldering. It is used in electronics, where it is used to connect electrical wiring and to connect electronic components to PCBs. Soldering is also used in plumbing to connect metal piping together with a water and gas tight bond.

Solder is a metal filler material that melts at a low temperature. For electrical connections it is usually composed of tin and lead in

FIGURE 3.36 Cord grip strain relief. (Courtesy of Thomas & Betts.)

Components and Hardware 113

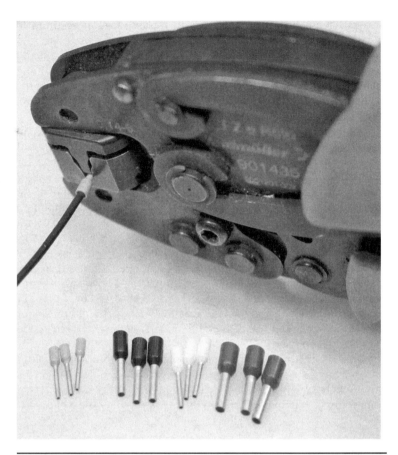

FIGURE 3.37 Ferrules.

various proportions, the most common being 63 percent tin and 37 percent lead. This proportion also has the benefit of being eutectic, meaning it passes directly from a liquid phase to a solid phase. This is important because metals that pass through an intermediate "plastic" phase are subject to cracking if disturbed while cooling.

Other alloys used for electrical connections are lead-silver for higher strength, tin-zinc or zinc aluminum for joining aluminum, and tin-silver and tin bismuth for other electronics. These alloys all melt at a lower temperature than the materials they are joining. This is the major difference between soldering and welding, which melts part of the workpiece. All of these alloys are known as soft solder, although silver solders are sometimes excepted from this classification.

The process of soldering involves melting the solder and flowing it into the joined wires or components. This process can be assisted by using a rosin water-based or "no-clean" type of "flux" to coat the joined pieces; solder flows to wherever the flux is applied. Many solders have a flux core that helps in this process. In addition to assisting the solder to flow, flux also helps clean the materials and prevent oxidation. When soldering stranded wires, solder is usually applied to the wire ends individually first; this is known as tinning the wires. If flux is applied to the strands beforehand or as part of the solder core, solder is drawn up into the strands by capillary action, called wicking.

Soldering by hand is done with a soldering iron, which is an electrically heated tool with an insulated handle and various different sizes of tip. Many of these have a temperature adjustment for different-size work. Figure 3.38 shows a soldering iron and roll of rosin-core solder. Often when soldering solid-state components, a clip-on "heat sink" is used between the wire lead and the component to prevent damage; proper temperature and tip size are important here also.

Soldering of components to PCBs on a production line is done by a process known as wave soldering. Components are adhesively attached to the board with leads extending through holes in the board and touching contact pads. The boards are then passed over pools of molten solder, which are vibrated, creating waves. This allows solder to contact the pads and leads without immersing the entire underside of the circuit board.

Another method of production soldering is to apply a solder powder and flux mixture in little clumps to the solder joint. This can then be melted with a heat lamp, hot air pencil, or most commonly in an oven. This method is called reflow soldering.

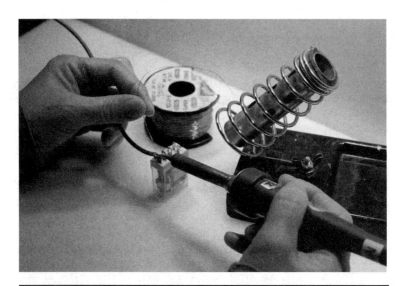

FIGURE 3.38 Solder and soldering iron.

Often a combination of wave, reflow, and hand soldering will be used on the same PCB.

3.5 Actuators and Movement

Actuators are used to move tooling on a machine, usually for the purpose of controlling the movement or positioning of a workpiece or sensor. They may be of a linear or rotating nature or a combination of both. Linear actuators are often used to generate a rotary movement by pushing a rotary pinned on an axis, or rotating devices such as motors can be used to generate linear motion via a belt or ball screw. Applications of these actuators are further discussed in section 3.7 of this primer.

A word on the nomenclature used in actuation. The words *home*, *advanced* or *extended*, *returned* or *retracted* are often used to describe the position of an actuator or its tooling. Great care must be taken to identify whether the designer is speaking of the tooling or the actuator itself. These positions can be the opposite of each other and cause physical rework and software changes if misinterpreted. It is preferable to refer to the position of the tooling generally since it is the most easily identified by maintenance or operators.

Descriptions such as "Tooling Raised" or "Pallet Stop Extended" can help reduce the ambiguity of generic movement labels for both electrical and mechanical designers.

3.5.1 Pneumatic and Hydraulic Actuators and Valves

Collectively the use of pneumatic and hydraulic energy is known as fluid power. The operation of actuators in fluid power applications is similar in the flow of liquids or gases through the systems; however, pneumatic systems use easily compressible air (or other inert gases) while hydraulic power is generated by the flow of much less compressible fluids, usually oil.

Pneumatic and hydraulic actuators may be linear or rotary in nature. Air cylinders generate a linear motion by injecting air through a port on one side or the other of a rounded piston surface inside a tubular housing. As air is injected through a valve into one end of the cylinder, the same valve releases air from the other side. A diagram of the internal configuration of an air cylinder is shown in Fig. 3.39. The end of the piston rod is threaded for attachment to various tooling pieces, such as a clevis or ball end.

Single acting cylinders use the force provided by air to move in one direction (usually out or "advanced") and a spring to return to the "home" or retracted position. Double acting cylinders use the air to move in both extend and retract directions. They have two ports to allow air in: one for the outward stroke and one for the return stroke. For a typical cylinder, the round piston face is attached to a rod extending through the end of the cylinder body. Some cylinders have

116 Chapter Three

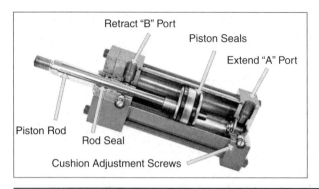

FIGURE 3.39 Pneumatic cylinder diagram.

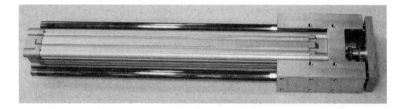

FIGURE 3.40 Guided air cylinder.

a rod attached to both faces of the piston face and extending out through both ends of the body. These are sometimes called double-ended or reciprocating cylinders.

Air cylinders are specified by their *bore*, or piston diameter, and their *stroke*, or how far the end of the shaft moves. Other specifications such as cushions to slow the last portion of motion, port sizes, and mounting method are also usually included in the part number. Sizes may be specified in both metric and standard measurements. Since stroke is specified in increments, stroke distances are sometimes limited by using shaft collars or limiting the movement of the tooling with stops. When this is done, the cushion may no longer be useful as it is at the farthest reach of the cylinder's stroke.

Figure 3.40 shows a very long stroke guided cylinder. These have bearings in a guide block that take the side load off the piston rod and ensure that force is applied linearly. Guide blocks can be ordered as a separate unit to mount a cylinder into.

Rodless air cylinders have no piston rod. They are actuators that use a mechanical or magnetic coupling to impart force, typically to a table or other body that moves along the length of the cylinder body but does not extend beyond it. These are also often called band cylinders. This is shown in Fig. 3.41.

Air cylinders are available in a variety of sizes and range from a small 2.5 mm diameter air cylinder, which might be used for picking

FIGURE 3.41 Rodless cylinder. (*Courtesy of SMC.*)

up a small electronic component, to 400 mm diameter air cylinders, which would impart enough force to lift a car. Some pneumatic cylinders reach 1000 mm in diameter and are used in place of hydraulic cylinders for special circumstances where leaking hydraulic oil might impose a hazard.

Pneumatic valves operate by using an electrically operated *solenoid* that shifts a spool inside the valve. This spool allows air to pass from an input port to an output port, also allowing air to escape from the exhaust side of the cylinder through the valve. The valves may be arranged in a variety of different ways, depending on the requirements of the application. Pneumatic valves are generally described by the number of ports in the valve body and the number of positions the spool may have. Like electrical circuits, they are also often specified as NC and NO, referring to their deenergized state. Examples of these are 2/2 and 3/2 valves. Most automated systems tend to use banks of 5/2 and 5/3 valves with open or blocked centers, depending on whether it is desirable to be able to move the actuator by hand in the deenergized state or not.

In addition to valves, fittings and devices like flow controls, pressure regulators, filters and a wide range of tubing and hoses are necessary to complete a pneumatic or hydraulic system. Accumulators and pressure intensifiers are also components sometimes used in pneumatic circuits. Table 3.1 shows pneumatic symbols for some of these valves and devices.

Hydraulic cylinders and actuators operate in a similar way to pneumatics except that they must be able to withstand much higher pressures and forces. More care must be taken to prevent the escape of fluids from the actuator also. Because of this, hydraulic actuators are more ruggedly built than typical pneumatic cylinders. External

Chapter Three

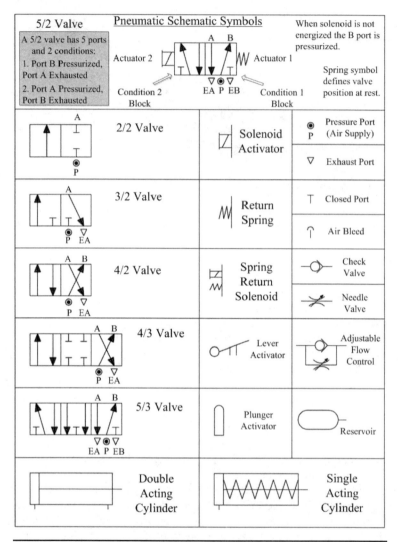

TABLE 3.1 Pneumatic Symbols

rods are often threaded into the end caps to help withstand the greater force exerted within the cylinder. Hydraulic cylinders are used in applications requiring great force, such as presses.

Unlike pneumatic systems, which are often supplied from a plantwide system, hydraulic systems have dedicated pumps. When oil is compressed, it generates heat, so the hydraulic fluid must also usually be cooled. Because of these extra components, hydraulic systems are much more expensive than pneumatic ones. Hybrid devices like air over oil actuators can sometimes help reduce the cost and complexity of hydraulic systems.

3.5.2 Electric Actuators

Electrically driven actuators are often used where air is not available or precision location is required. Though typically more costly than an air cylinder, they are less expensive and complex than a hydraulic system. Electric actuators are often servomotor driven and ball screw or belt based. They can be found in many of the same packaged configurations as air cylinders.

Small magnetic solenoids may also be used to extend a rod a short distance; these consist of a coil of wire wrapped around a bushing with a metal rod inside. A well-known example of this type of actuator is those used in pinball machines for the flippers and bumpers. The spool in solenoid valves uses this same principle.

3.5.3 Motion Control

Motion control is often considered an entire subchapter within the field of automation. Motion control differs from standard discrete controls such as pneumatic cylinders, conveyors, and the like because the positions and velocities are both controlled by analog or digitally converted analog methods. This is accomplished by the use of hydraulic or pneumatic proportional valves, linear actuators, or electric motors, usually servos. Stepper motors are also a common component in small motion control systems, especially when feedback may not be economical. Motion control is used extensively in packaging, printing, textile, semiconductor, and assembly industries. It also forms the basis of robotics and CNC machine tools.

The basic architecture of a motion control system consists of:

- A motion controller to generate the desired output or motion profile. Movement is based on programmed set points and closing a position or velocity feedback loop.

- A drive or amplifier to transform the control signal from the motion controller into a higher-power electrical control current or voltage. This is what is applied to the actuator and actually makes it move.

- An actuator such as a hydraulic or air cylinder, linear actuator, or electric motor for output motion.

- One or more feedback sensors, such as optical encoders, resolvers, or Hall effect devices. These return the position or velocity of the actuator to the motion controller in order to close the position or velocity control loops. Newer "intelligent" drives can close the position and velocity loops internally, resulting in more accurate control.

- Mechanical components to transform the movement of the actuator into the desired motion. Examples are gears, shafting, ball screws, belts, linkages, and linear and rotational bearings.

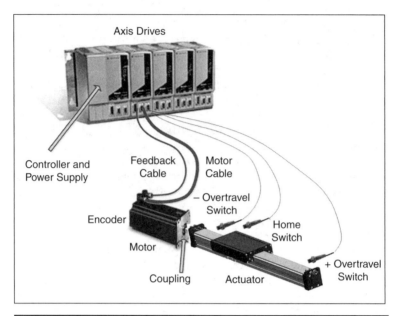

Figure 3.42 Motion control system.

Figure 3.42 shows the physical arrangement of a motion control system.

A stand-alone motion control axis is common when positioning; however, there are times when motions must be coordinated closely. This requires tight synchronization between axes. Robotics is an example of a coordinated motion system working together. Prior to the development of fast open communication interfaces in the early 1990s, the only open method of coordinated motion was analog control being brought back to the controller in the form of encoder, resolver, and other analog methods such as 4 to 20 mA and 0 to 10 V signals. The first open digital automation bus to satisfy the requirements of coordinated motion control was Sercos (www.sercos.com). This is an international standard that closes the servo feedback loop in the drive rather than in the motion controller. This arrangement reduces the computational load on the controller, allowing more axes to be controlled at once. Since the development of Sercos, other interfaces have been developed for this purpose, including ProfiNet IRT, CANopen, EtherNet PowerLink, and EtherCAT.

Besides the common control functions of velocity and position control, there are several other functions that may be considered. Since torque feedback can be determined from the current and velocity of the servo, pressure or force control is another function of a servo actuator. Electronic gearing can be used to link two or more axes together in a master/slave relationship. Cam profiling where one axis follows the motion of a master axis is an example of this.

More detailed profiles, such as trapezoidal moves or S-curves, can also be computed by a motion controller enhancing standard positional moves. This helps to eliminate acceleration or deceleration impacts such as "jerk."

One of the best online resources for motion control theory and components is Motion Control Resource (www.motioncontrolresource.com). This site is amazingly free of advertisements and contains links to many major motion control component manufacturers and distributors.

3.6 AC and DC Motors

An electric machine is a link between an electrical system and a mechanical system. The process of converting energy from one of these forms to the other is electromechanical energy conversion. In these machines, the process is reversible. If the conversion is from mechanical to electrical, the machine is acting as a generator, and if the conversion is from electrical to mechanical, the machine is acting as a motor.

Three types of electrical machines are used extensively for electromechanical energy conversion: DC, induction, and synchronous motors. Other types of motors are permanent magnet (PM), hysteresis, and stepper motors. Conversion from electrical to mechanical energy is based on two electromagnetic principles: when a conductor moves within a magnetic field, voltage is induced in the conductor; simultaneously, when a current-carrying conductor is placed in a magnetic field, the conductor experiences a mechanical force. In a motor, an electrical system makes current flow through conductors placed in the magnetic field and a force is exerted on each conductor. If the conductors are placed on a structure that is free to rotate, an electromagnetic torque is produced, making the structure rotate. This rotating structure is called a *rotor*. The part of the machine that does not move and provides the magnetic force is called the *stator*. Usually this is the outer frame of the machine or motor with the exception of special cases such as powered rollers.

Both stator and rotor are made of ferromagnetic (iron-rich) materials. The iron core is used to maximize the coupling between the coils of wire, increasing the magnetic flux density in the motor and therefore allowing its size to be reduced. In most motors, slots are cut on the inner periphery of the stator and outer periphery of the rotor and conductors are placed in the slots. If a time-varying electrical signal is placed on the stator or rotor (or both), it will cause a mechanical torque to be exerted by the rotor. The conductors placed in the slots are interconnected to form windings; the winding through which the current is passed to create the major source of magnetic flux is called the field winding, although in some motors the main source of magnetic flux is a PM.

Electric motors are used in many different applications of automated systems, from blowers, pumps, and fans to conveyors, robotics, and actuators. They may be powered by AC supplied from a power grid within the plant or a motor drive, or DC from batteries or a converter. Motors may be classified by their construction method, their source of power, or their application and the type of motion they provide. In the industrial field they are generally standardized as to size and horsepower or wattage range.

3.6.1 AC Motors

A typical AC motor consists of two parts: a stator having coils supplied with AC current to produce a rotating magnetic field and an inside rotor attached to an output shaft. The rotor is provided a torque by the rotating field that is generated by the alternating current.

AC motors often include designations relating to their physical construction such as TE (totally enclosed), FC (fan cooled), and PM. Other information, such as frame size, also describes motors physically, including mounting options, sealing methods, and shaft sizes. A good motor catalog will describe these options well.

Synchronous Motors

A synchronous motor is an AC machine with a rotor that rotates at the same speed as the alternating current that is applied. This is accomplished by exciting the rotor's field winding with a direct current. When the rotor rotates, voltage is induced in the armature winding of the stator; this produces a revolving magnetic field whose speed is the same as the speed of the rotor. Unlike an induction motor, a synchronous motor has zero "slip" while operating at speed.

Slip rings and brushes are used to conduct current to the rotor. The rotor poles connect to each other and move at the same speed; hence, the name *synchronous motor*. Synchronous motors are used mainly in applications where a constant speed is desired and are not as common in industrial applications as induction motors.

One problem with synchronous motors is that they are not self-starting. If an AC voltage is applied to the stator terminals and the rotor is excited with a field current, the motor will simply vibrate. This is because as the AC voltage is applied it is immediately rotating the stator field at 60 Hz, which is too fast for the rotor poles to catch up to. For this reason synchronous motors have to be started by either using a variable frequency supply (such as a drive) or starting the machine as an inductive motor. If a drive is not used, an extra winding can be used called a "damper" winding. In this instance, the field winding is not excited by DC but is shunted by a resistance. Current is induced in the damper winding, producing a torque; as the motor approaches synchronous speed, the DC voltage is applied to the rotor and the motor will lock onto the stator field.

Three-Phase AC Synchronous Motors

The stator of a three-phase synchronous motor has a distributed winding called the armature winding. It is connected to the AC supply and is designed for high voltage and current. DC is then applied to the rotor coils of the motor through slip rings and brushes from a separate source. This creates a continuous field, and the rotor will then rotate synchronously with the alternating current applied to the stator.

Synchronous motors can be further divided by two different construction types: high-speed motors with cylindrical rotors and low-speed motors with salient pole rotors. The nonsalient pole or cylindrical motor has one distributed winding and a uniform air gap between the rotor and stator. The rotor is generally long and has a small diameter. These motors are often used in generators.

Salient pole motors have concentrated windings on the motor poles and a nonuniform air gap. The rotors are shorter and have a greater diameter than cylindrical rotor synchronous motors. Salient pole motors are often used to drive pumps or mixers.

One use for a synchronous motor is its use in a *power factor correction* scheme; these are referred to as synchronous condensers. This method uses a feature of the motor where it consumes power at a leading power factor when its rotor is overexcited. It appears to the supply to be a capacitor, and can then be used to correct the lagging power factor that is usually presented to the electric supply by inductive loads. Since factories are charged extra for their electricity consumption if the power factor is too low, this can help correct a plant's power profile. The excitation is adjusted until a near unity power factor is obtained (often automatically). Motors used for this purpose are easily identified as they have no shaft extensions.

Single-Phase AC Synchronous Motors

Small single-phase AC motors can also be designed with PM rotors. Since the rotors in these motors do not require any induced current, they do not slip backward against the stator frequency; instead, they rotate synchronously. Because they are very accurately synchronized with the applied frequency, which is carefully regulated at the power plant, these motors are often used to power mechanical clocks, chart recorders, or anything else that requires a precise speed.

Hysteresis synchronous motors use the hysteresis property of magnetic materials to produce torque. The rotor is a smooth cylinder of a magnetic alloy that stays magnetized but can be demagnetized fairly easily as well as remagnetized with poles in a new location. The stator windings are distributed to produce a sinusoidal magnetic flux. Because of the hysteresis of the magnetized rotor, it tends to lag behind the rotating field. This creates a constant torque up to the synchronous speed, a useful feature for some applications. A hysteresis motor is quiet and smooth running; however, it is more expensive than a reluctance motor of the same rating.

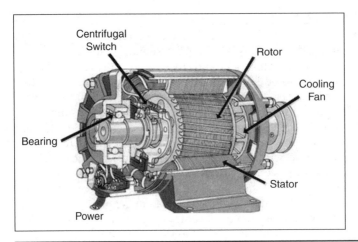

FIGURE 3.43 Single-phase "squirrel cage" motor.

A *reluctance motor* has a single-phase distributed stator winding and a cage-type rotor, often called a "squirrel cage." This is a cylindrical-shaped rotor with bars spaced around the periphery. In a reluctance motor, some of these teeth are removed. The stator of a single-phase reluctance motor has a main winding and an auxiliary starting winding. When the stator is connected to a single-phase supply, the motor starts as an induction motor. A centrifugal switch is then used to disconnect the auxiliary winding at about 75 percent of the synchronous speed. The motor continues to gain speed until it is synchronized with the rotating field. Reluctance motors are generally several times larger than an equivalent horsepower motor with DC excitation; however, because it has no slip rings, brushes, or field winding, it is low cost and fairly maintenance-free. A single-phase squirrel cage motor is shown in Fig. 3.43.

Asynchronous Motors

Induction motors are the most rugged and widely used motor for industrial applications. An induction motor has a stator and rotor with a uniform air gap between their windings. The rotor is mounted on bearings and is made of laminated sheets of ferromagnetic metal with slots cut on the outer surface. The rotor winding may be of the squirrel cage type or the wound rotor type. The stator is also made of laminations of high-grade sheet steel with distributed windings. In induction motors, alternating current is applied to both the stator and rotor windings.

Three-Phase AC Induction Motors

Windings of both the stator and the rotor of a three-phase motor are distributed over several slots in the laminated sheets. Terminals of the rotor windings are connected to three slip rings; using stationary

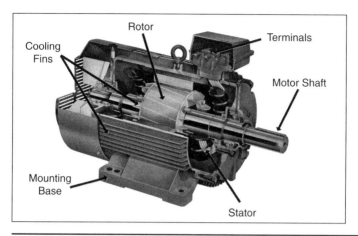

FIGURE 3.44   Three-phase AC induction motor.

brushes, the rotor can then be connected to an external circuit. Power applied to the three-phase windings of the stator and rotor produce rotating fields 120° apart electrically, as shown in the waveform for three-phase power in Chap. 2. A cutaway diagram of a three-phase induction motor is shown in Fig. 3.44.

Current is induced in the rotor by the rotating fields of the stator. As the rotor rotates, the relative speed of the rotor and fields decreases as the motor speeds up. If the rotor speed were to reach the rotating field speed, the rotor would provide no torque. The difference between the rotor speed and the synchronous speed is called *slip*. When loaded, standard motors have between 2 and 3 percent slip; a three-phase 60-Hz motor typically runs at 1725 to 1750 rpm as opposed to a calculated speed of 1800 rpm.

Induction motors are the most commonly used AC motors in industrial automation and are produced in standard frame sizes up to about 500 kW or 670 horsepower. This makes them easily interchangeable, though European and North American standards are different.

Single-Phase AC Induction Motors

Most single-phase induction motors have squirrel cage rotors and a single-phase distributed stator winding. Some single-phase induction motors use a wound rotor, but these are far less common. The squirrel cage motor takes its name from its shape—a ring at either end of the rotor connected by bars running along its length, forming a cage shape.

Single-phase induction motors are classified by the methods used to start them. Some common types are resistance-start or split-phase, capacitor start, capacitor run, and shaded pole.

The *split-phase* induction motor has a main winding and an auxiliary winding on the stator. The auxiliary winding is used for starting as described in the reluctance synchronous motor. The two windings are placed 90 electrical degrees apart and the currents of the two windings are therefore phase shifted from each other. This produces a starting torque; the auxiliary winding can then be removed from the circuit using a centrifugal switch as described previously.

If a capacitor is placed in series with the auxiliary winding, a greater phase angle is created, creating a higher starting torque. This method of starting is known as a *capacitor start* motor. The cost of this motor is slightly higher than that of the split-phase type, though because the circuit is only used for starting an inexpensive AC electrolytic capacitor can be used.

In a *capacitor run* motor, the starting capacitor and auxiliary winding are not removed from the circuit while at full speed. This requires a different kind of capacitor, usually an AC paper-oil type. Although the capacitor is more expensive than the electrolytic type, the centrifugal switch is removed, reducing the cost. Starting torque is not as high as that of the capacitor start type; however, the motor is quieter running.

If both optimum starting and optimum running torque are desired, a combination starting method called *capacitor-start capacitor-run* can be used. This places an electrolytic capacitor in series with the auxiliary winding and a smaller value paper-oil-type capacitor in series with the main winding. This is a more expensive motor than the others; however, it provides the best performance.

Shaded pole motors use the salient pole construction method described previously in the synchronous motor. The main winding is wound on the salient poles, but a short-circuited copper turn is placed between the main coil and the rotor, "shading" the magnetic flux as it rotates. This creates a small starting torque. This method is used in low torque applications, such as fans or small devices.

A *resistance start* motor is a split-phase induction motor with a resistance inserted in series with the start-up winding, creating a starting torque. The resistance provides assistance in the starting and initial direction of rotation without producing excess current. Starting torque in a resistance start motor is higher than that of a shaded pole or capacitor run motor, but not as high as a capacitor start.

3.6.2 DC Motors

A DC motor places the armature winding on the rotor and the field windings on the stator, which is the opposite of the AC motors described previously. It is designed to run on DC power, though it alternates the direction of current flow in the windings through commutation. The stator has salient or projecting poles excited by one or more field windings; these produce a magnetic field that is symmetrical around the pole axis, also called the field or direct axis. The voltage induced in the armature winding alternates by using a commutator-brush combination as a mechanical rectifier. Alternatively, a brushless DC motor uses an external electronic switch synchronized to the position of the rotor.

The field and armature windings can be connected in a variety of ways to provide different performance characteristics. The field windings can be connected in series, in shunt (parallel with the armature), or as a combination of both, called a compound motor. DC motors can also have a PM.

Brushed DC Motors

The field winding is placed on the stator to excite the field poles and the armature winding is placed on the rotor. The commutator consists of a split ring connected to each end of the rotor windings. DC voltage is then applied to the brushes; as the rotor turns, the brushes alternately contact the different halves of the ring, changing the direction of the current flow and thereby creating an alternating field. This field never fully aligns with the salient poles of the stator, which keeps the rotor moving.

More than one set of rings and poles can be and often are used in larger DC motors. The distance between the centers of adjacent poles is known as pole pitch, while the difference between the two sides of the coil is called coil pitch. If the coil pitch and pole pitch are equal, it is called a full-pitch coil. A coil pitch that is less than a pole pitch is known as a short pitch or fractional pitch coil. AC motors often have short pitch coils, while DC motors have full-pitch coils. Figure 3.45 illustrates the construction of a brushed DC motor.

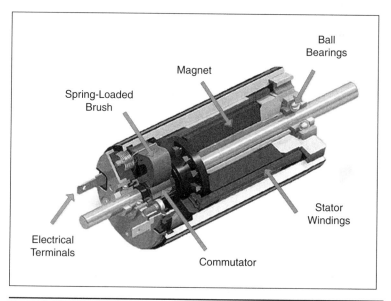

FIGURE 3.45 Brushed DC motor.

Disadvantages of Brushes

Because brushes constantly wear as they press against the commutator rings, they eventually have to be replaced. Brushes also create sparks as they cross the insulating gaps in the commutator. At high speeds the brushes have a harder time maintaining contact with the commutator; this also creates sparking. Sparking can pit the commutator surface, creating irregularities and making the contacts of the brushes bounce, which causes even more sparking. This can overheat and eventually destroy the commutator and brushes. Brushed DC motors also create quite a bit of electrical noise because of this sparking, and maximum speed is limited.

Many of the problems created by the brushes are eliminated in brushless motors, which last longer and are more efficient in their use of energy.

Some of the problems of the brushed DC motor are eliminated in the brushless design. In this motor, the mechanical "rotating switch" or commutator brush gear assembly is replaced by an external electronic switch synchronized to the rotor's position. Brushless motors are typically 85 to 90 percent efficient or more (higher efficiency for a brushless electric motor of up to 96.5 percent was reported by researchers at the Tokai University in Japan in 2009), whereas DC motors with brushes are typically 75 to 80 percent efficient.

Brushless DC Motors

The brushless DC motor replaces the brushes and commutator with an electronically alternating pulse that is synchronized to the position of the rotor. Hall effect sensors are used to sense the position of PMs on the rotor and the driving coils are activated sequentially. Coils are usually arranged in groups of three, acting very similarly to a three-phase synchronous motor.

Another method of sensing rotor position is by detecting the back-EMF in the inactivated driving coils. This allows the drive electronics to sense both speed and position of the motor. These motors are often used in applications where very accurate speed control is required.

Brushless DC motors last much longer than those with brushes and run cooler than AC motors. They are very quiet from an electrical noise standpoint as well as audibly. Since they do not create sparks like motors with brushes, they are better suited to chemical or explosive environments.

Coreless or Ironless DC Motors

A motor capable of very rapid acceleration is the coreless or ironless motor. This motor makes use of a very lightweight rotor by making it almost entirely of the windings themselves with no steel or ferromagnetic material in the rotor. This method of construction can be used for brush and commutator or brushless motors. The rotor can either be placed inside the stator magnets or form a cylindrical basket shape outside the stator. Windings for these rotors are often encapsulated in epoxy for physical stability. These types of motors are also typically rather small. They also tend to generate quite a bit of heat since there is no metal to act as a heat sink; this often necessitates an additional cooling method, such as forcing air over the rotor windings.

Universal Motors and Series Wound DC Motors

DC motors with the field and armature windings placed in series allow the motor to run on either AC or DC power. These motors are called universal or series wound motors. Though very flexible as far as power usage, they have several disadvantages when comparing them with standard AC or DC varieties.

As a universal motor increases its speed, its torque output decreases, making it impractical for high-speed high torque applications. Without a load attached, these motors also tend to "run away," potentially damaging the motor. A permanent load such as a cooling fan is often attached to the shaft to limit this problem. The high starting torque can be useful in some starting applications.

Universal motors operate better using DC than AC and are best for intermittent use. Accurate speed control can also be problematic.

3.6.3 Linear Motors

Linear motors operate in a similar manner to standard electric motors except that the rotor and stator are placed next to each other in a linear fashion, or "unrolled." Generally linear motors are classified as either low or high acceleration. AC linear induction motors (LIMs) are used for high acceleration applications. Typically they use a powered stator winding with a conducting plate as the rotor carrying the load.

Linear synchronous motors (LSMs) are used for larger motors requiring high speed or high torque. They also use a powered stator winding but use an array of alternating pole magnets mounted to the load-bearing frame as a rotor. These motors have a lower acceleration than the LIM type.

3.6.4 Servomotors and Stepper Motors

Servomotors are specially designed and built for use in feedback control systems. This requires a high speed of response, which servomotors achieve by having a low rotor inertia. Servomotors are therefore smaller in diameter and longer than typical AC and DC motor form factors. They must often operate at low or zero speed, which makes them typically larger than conventional motors with a similar power rating. Peak torque values are often 3x continuous torque ratings, but may be as high as 10x.

Servo power ratings can range from a fraction of a watt to several hundred watts. Within a specific power range, different inertias may also be specified by some motor manufacturers. They are used in a wide variety of industrial applications, such as robots, machine tools, positioning systems, and process control. Both AC and DC servomotors are used in industry.

Brushless servomotors often use sinusoidal commutation to produce smooth motion at lower speeds. If the more traditional trapezoidal or "six-step" DC commutation method is used, motors tend to "cog" or produce a jerky motion at low speed, partially because of the low inertia of servomotors. Motors rotate because of the torque produced by the interacting magnetic fields of the rotor and stator. The torque is proportional to the magnitudes of the fields multiplied by the sine of the angle between them. Maximum torque is produced when the rotor and stator angles are at 90°. Torque can then be controlled by varying the angle between the two waveforms. To detect the relative positions of the rotor and stator, a commutation encoder can be used to find the phase angles relative to each other. These are incremental encoders with additional tracks for regulating motor commutation.

Servomotors are driven by servo drives that provide precise velocity, torque, and position control by using encoder, resolver, and/or current signals that comprise the feedback components of a servomechanism. Additional components of a servomechanism actuator are a home switch to establish a reference position and overtravel switches to prevent actuator or tooling damage.

DC Servos

DC servomotors may be separately excited or PM DC motors. The principle of operation is the same as described in the DC motor section 3.6.2 previously. They are normally controlled by varying the armature voltage, which has a large resistance, ensuring that the torque-speed ratio is linear. The torque response is very fast in these motors, making them ideal for quick changes in position or speed.

AC Servos

AC servos are robust in construction and have a lower inertia than DC servomotors; however, they are nonlinear in their torque-speed response. They also have lower torque capability than DC servos of a similar size.

Most AC servos are two-phase squirrel cage–type motors. The stator has two distributed windings displaced 90° electrically. One winding, the reference or fixed phase winding, is connected to a constant voltage source. The other winding is called the control phase and is supplied with a variable voltage at the same frequency as the reference phase. For industrial applications, the frequency is usually 60 Hz. The control phase voltage is supplied from a servo amplifier, which controls rotation direction by shifting the phase plus or minus 90° from the reference voltage.

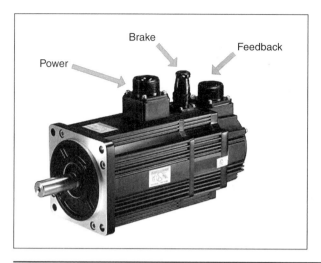

FIGURE 3.46 AC servomotor.

Figure 3.46 shows a typical AC servomotor with its electrical cable connections. A gearbox is often bolted to the motor flange.

The squirrel cage rotor has a high resistance like the DC rotor windings; varying this resistance provides different torque-speed characteristics. Lowering the resistance decreases the torque at low speed and increases it at higher speeds, making the curve very nonlinear. This is not desirable in control systems.

Two-phase AC servomotors are built as high-speed low torque actuators and are usually geared down severely to achieve the desired result. Typical speeds of these motors are 3000 to 5000 rpm.

Stepper Motors

A stepper motor is a DC motor that rotates a specific number of degrees based on its construction, that is, number of poles. It converts digital pulse inputs to shaft rotation; a train of pulses is made to turn the motor shaft by steps. This allows the position to be controlled precisely without a feedback mechanism. Typical resolutions of commercially available stepper motors range from a few steps per revolution to as many as 400. They can follow signals of up to 1200 pulses per second and may be rated up to several horsepower.

There are several different types of stepper motors, including single and multiple stack variable reluctance motors and PM types. Variable reluctance motors operate by exciting the poles of the stator, causing the rotor to align itself with the magnetic field. The poles may be energized in combinations, allowing the rotor to line up between stator poles as well as directly with them. Multiple stack versions arrange the poles in several levels or "stacks," allowing finer resolution positioning by phasing from stack to stack.

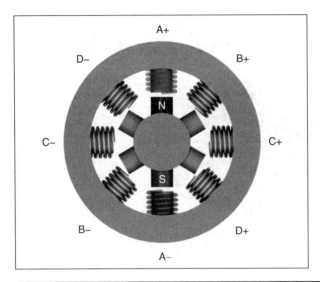

Figure 3.47 Stepper motor diagram.

PM steppers use magnets for the rotor poles. They have a higher inertia than variable reluctance motors and therefore cannot accelerate as fast; however, they produce more torque per ampere of stator current. Figure 3.47 shows a four-pole stepper arrangement with PMs; the A, B, C, and D poles are energized in sequence in one polarity after which the polarities are reversed to achieve eight positions per revolution.

Hybrid stepper motors use a combination of variable reluctance and PM motor techniques. This provides maximum power in a small package size. Hybrid stepper motors are probably the most commonly used type of stepper in industrial automation.

Though steppers can be a lower-cost alternative to servos for positioning applications since feedback is not required, stepper motors do not provide nearly as much torque as servomotors, especially at higher speeds.

Command signals for stepper motors are usually low power logic circuits using TTL or CMOS transistors, power amplification stages are placed between the pulse train generators and the motors.

3.6.5 Variable Frequency Drives

Variable frequency drives (VFDs) are solid-state power converters. They first convert an incoming AC voltage into DC, then reconstruct an AC waveform by switching the DC power rapidly at the desired frequency and voltage to approximate a sinusoidal signal. The rectifier that converts the incoming voltage to DC is usually a three-phase full wave bridge; single-phase power may also be used for smaller VFDs. Figure 3.48 is a diagram of this system.

In order to deliver a consistent torque value while varying speed, the applied voltage must be adjusted proportionally with the

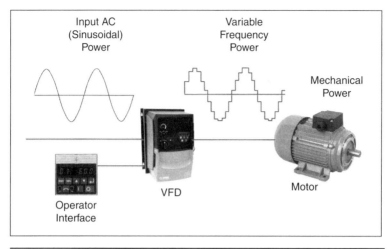

FIGURE 3.48 VFD system.

frequency. If a motor is rated for 480VAC at 60 Hz, the voltage must be reduced to 240VAC for 30 Hz, 120VAC for 15 Hz, and so on. This is sometimes called volts per hertz control. Additional methods such as vector control and direct torque control allow the magnetic flux and mechanical torque of the motor to be controlled more precisely.

The stage that converts the DC back into a sinusoidal form is known as an inverter circuit. This circuit usually uses pulse-width modulation (PWM) to adjust both output voltage frequency and voltage as required. This is illustrated in Fig. 3.49.

Newer drives often use special transistors called IGBTs, or Insulated Gate Bipolar Transistors. These are electronic switches that operate over a wide current range, have high efficiency and fast switching, making them ideal for PWM.

A microprocessor is used to control the operation of the VFD. Typically there are a range of parameters that can be set to control the operation of the drive: acceleration and deceleration, maximum speed and velocity set points, and peak current are some of the more common values. Digital I/O connections for start/stop, alarms, and speed preset selection are also common. These may be hardwired or communications based. Analog values may also be interfaced with the drive physically as in a 0 to 10 V or 4 to 20 mA signal or via mapping communication values from a controller.

An OIT may also be mounted on the front of the drive for setting parameters and viewing operational data, such as current or speed. These may be built into the drive or removable so that it can be shared between VFDs. Like servo systems, VFDs can also be used with feedback devices such as encoders and resolvers to improve control; however, typically a controller such as a PLC or DCS is used as an intermediary between the device and the drive.

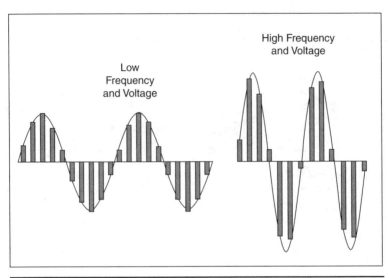

FIGURE 3.49 Pulse width modulation.

VFDs can be operated at speeds above the speed listed on the nameplate of the motor, depending on the application. At ranges above 150 percent, it is usually recommended that a gearbox be used. Another consideration when planning a system using a VFD is the distance of the motor from the drive. At distances over 150 ft or so, a phenomenon called reflected wave can occur because of the rapid switching of the transistors. This can cause high voltages to be present in the cabling and motor. There are a number of ways to mitigate this, including filters and using inverter duty motors, but ideally the drive should be located relatively close to the motor.

3.7 Mechanisms and Machine Elements

Mechanisms or machine elements make up the basic components of mechanical systems. The primary purpose of a mechanism is to transfer or transform force from one form or direction into another. The most basic elements of mechanisms were described as "simple machines" by Renaissance scientists and included the following:

- Lever
- Wheel and axle
- Pulley
- Inclined plane
- Wedge
- Screw

Gears and cams were developed as an offshoot of several of these elements and are also important types of mechanisms. The classical concept of decomposing machines into these simple elements still has relevance today, though there are elements that do not fall directly into these categories. Mechanisms and simple machines can be thought of as the building blocks of more complex machines.

Machine elements include components that allow power to be transmitted from one mechanism to another. Elements such as bearings, couplings, clutches, brakes, belts, and chains are examples of components that facilitate movement.

3.7.1 Cam-Driven Devices

One method of translating rotary motion into linear motion is the use of a cam on a rotating shaft. By offsetting the center of a round or oval disc on the shaft, the cam surface will vary in distance from the shaft center. This can be used to drive a shaft linearly; the linear shaft is often called a follower. Springs are used to keep the end of the follower in contact with the cam as it rotates. Figure 3.50 shows a cam and follower arrangement.

Cams are used extensively in line shaft–type applications. They allow stations to be synchronized with a master motor and run at higher speeds than standard asynchronous applications. The downside is that they are more difficult to modify movement profiles as the cams must be machined or replaced.

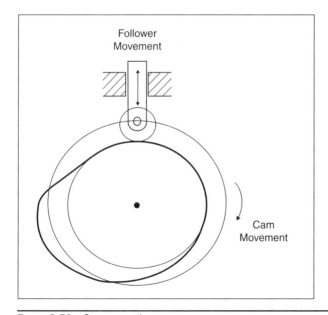

Figure 3.50 Cam operation.

FIGURE 3.51
Ratchet and pawl.

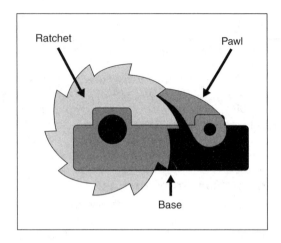

3.7.2 Ratchet and Pawl Systems

A ratchet is a mechanism that allows a linear or rotary movement in one direction only. Movement in the opposite direction is prevented by a spring-loaded pawl that engages teeth on the ratchet as it turns. The teeth are angled such that the pawl cannot be forced out of the slots between the ratchet teeth when the ratchet is reversed. Figure 3.51 illustrates this arrangement.

Ratchet and pawl systems are used in lifting mechanisms such as jacks and winding mechanisms, and even plastic cable ties. Ratchet gearing may be used to transmit intermittent motion or to simply prevent the reverse movement of the gear.

3.7.3 Gearing and Gear Reduction

Gearing is used to translate rotary motion from one speed, direction, or force into another. A gear is a mechanism, usually round, that has teeth that engage with another toothed device. The mating interface between the two machine parts is called a spline.

Gears can be combined into "trains," which can change speeds and therefore torque outputs incrementally rather than all at once. They may have teeth on the outside or inside circumference or a combination of both. Gear tooth profiles are almost always slightly curved in a shape called an involute curve. This curve is based on the diameter of the gear and is important in keeping gear movement and interfacing smooth and consistent.

Gear reduction is the process of converting a high-speed low torque component, such as a servomotor, into a lower-speed higher torque output without creating excessive backlash. This is usually done with a self-contained gearbox, which also may change the shaft or rotation direction.

The simplest type of gear is the cylindrical *spur gear*. Spur gears only mesh if the gear axles are parallel with each other. There are two types of spur gears: internal and external. External spur gears transmit drive between parallel shafts, making them rotate in opposite directions. This is illustrated in Fig. 3.52. They work well at moderate speeds but can be noisy at higher speeds. Internal spur gears are spur gear arrangements that transmit motion to shafts rotating in the same direction.

Helical gears have teeth cut at an angle to the axis. As with the spur gear arrangements, they can be arranged in an internal or external manner. Unlike spur gears, however, helical gears can also mesh on nonparallel axes. Whereas spur gears must be parallel but produce thrust only perpendicular to the load, helical gears produce axial thrust when arranged with parallel shafts. They may be arranged with the shafts at an angle to each other, even completely perpendicular— his is known as a crossed helical gear arrangement. Figure 3.53 illustrates a set of helical gears at a slight angle to each other.

Helical gears can take more torque than spur gears and are often used in higher-speed applications. The higher torque causes more axial thrust, however, which can be mitigated by using double helical gears. These are the equivalent of two mirrored helical gears stacked on top of each other and are sometimes called herringbone gears. Because of the more gradual engagement of the teeth, helical gears are quieter than spur gears. The more complex profile makes double helical gears more expensive than helical or spur gears.

Bevel gears are conical gears with the teeth cut at an angle to the shaft. They are designed to connect two shafts on intersecting axes, as shown in Fig. 3.54. Straight bevel gears have teeth that are cut radially toward the apex of the conical section. Spiral bevel gears have curved oblique teeth that reduce noise and improve the smoothness of meshing between gears, similar to the effect of the helical gear.

Figure 3.52 Spur gears.

Figure 3.53 Helical gears.

Figure 3.54
Bevel gear.

Hypoid bevel gears are a combination of spiral bevel gears and worm gears. The axes of these gears do not intersect. The distance between the axes is called the offset. Hypoid bevel gears allow a greater gear ratio than regular bevel gears, making them a good choice for gear reduction in mechanical differentials.

Crown gears have teeth that project perpendicular to the plane of the gear or parallel with the shaft. They are considered to be part of the bevel gear grouping and are sometimes called contrate gears. They may be meshed with other bevel-type gears or spur gears.

A *worm gear* is used to transmit motion at a right angle to its shaft. They have line tooth contact and are often meshed with disk-type gears, sometimes called the wheel or worm wheel. A worm gear resembles a screw and may have one or more toothed tracks running around it, as shown in Fig. 3.55. Because the pitch of the worm can be made quite small, high ratios of gear reduction can be achieved; however, this is at the cost of efficiency. When a worm and gear combination is used, the worm can always drive the gear; however, the reverse is not always true. If the ratio is high enough, the teeth may lock together because the force exerted by the wheel cannot overcome the friction of the worm gear. This can be an advantage if it is preferable to hold a worm-driven object in position against the force of gravity.

Worm gearing can be divided into two general categories—fine and coarse pitch gearing. The main purpose of fine pitch gearing is to transmit motion rather than power, while the opposite is true of coarse pitch.

A *rack* is a linear toothed rod or bar that usually engages with a round gear called a pinion gear. This is a common method of converting rotational motion into linear motion and vice versa. As with other gear types, teeth may be cut straight or at an angle to the axis of motion. A rack is often used in gear theory as a gear of infinite radius.

FIGURE 3.55
Worm and gear combination.

A rack and pinion system is an excellent method of moving a linear axis rapidly over a long distance. The rack is usually the fixed component, and the pinion gear is rotated from the traveling part of the system, which is guided by linear bearings. Figure 3.56 shows an industrial rack and pinion.

Planetary or epicyclic gearing is a method of combining gears in such a way that one or more of the gear axes is movable, usually one rotating around another. There are various arrangements that accomplish this using bevel or spur-type gears. Epicyclic gearing is a very compact method of achieving gear reduction and is often used in servo gearboxes. Figure 3.57 shows a cutaway of a planetary gearbox.

Figure 3.56 Rack and pinion gear.

Figure 3.57 Planetary gearbox. (*Courtesy of JVL.*)

3.7.4 Bearings and Pulleys

Bearings allow sliding or rolling contact between two or more parts. They fall into three general categories based on their purpose: radial bearings that support rotating shafts or journals, thrust bearings that support axial loads on rotating elements, and guide bearings that support and guide moving elements in a straight line. Bearings are also often described by the principle of operation or the direction of the applied load.

Bearings that provide sliding contact are known as plain bearings. Relative motion between the parts of plain bearings may be of the lubricated type, either a hydrodynamic interface (a wedge or film buildup of lubricant is produced by the bearing surface) or hydrostatic interface (a lubricant is introduced into the mating surfaces under pressure). The motion interface may also be unlubricated with a material such as nylon, brass, or Teflon. Plain bearings are also known as bushings when they operate on a shaft.

Rolling contact–type bearings use rolling elements, such as balls or rollers, in place of lubricants or direct contact. Figure 3.58 shows a cutaway of a roller bearing with cylindrical rollers. Roller bearings generally have a much lower friction coefficient than plain bearings and therefore have less energy loss. They also generally hold tighter tolerances and are consequently more precise. Normally, the rolling elements and the races they ride in are hardened to reduce wear. Roller-type bearings are also generally shielded or sealed to reduce the chance of contaminants entering the bearing races.

The use of linear roller bearings and rails is one of the most common guiding methods for linear movement. They often support ball screw or worm gear–driven axes in motion control applications. A rail with two bearing blocks is shown in Fig. 3.59.

FIGURE 3.58
Cylindrical roller bearing.

Figure 3.59 Linear bearings and rail.

An *air bearing* is a pneumatic device that uses a film of air between surfaces. They are often used in moving heavy loads on a floor surface, similar to a hovercraft or air hockey table. Rotary, spindle, and slide air bearings provide almost no resistance to motion and are very precise. Air bearings may be externally pressurized with a continuous flow or generate a film of air from the relative motion of the two surfaces.

A *pulley*, sometimes called a sheave, is a wheel or drum mounted on an axle. It generally has a groove or channel between two flanges that carries a belt, chain, or cable. Pulleys can change the direction or speed of a motion, much in the same manner as gearing. Pulleys of different dimensions transfer speed changes in proportion to the diameters or circumferences of the pulleys. For example, if a pair of pulleys has a diameter ratio of 2 to 1, the speed will be increased proportionally.

Pulleys are often used with flexible belts in industrial applications. It is common to use a steel-reinforced toothed belt in a belt-driven actuator to provide linear motion. In this case, the pulley will also have grooves in the surface parallel to the shaft, as shown in Fig. 3.60. These are called positive drive belts. Belt and pulley combinations without positive drive are known as friction drives.

The number of degrees that the belt is in contact with the pulley is called the wrap angle. Belt and pulley combinations should be selected such that the wrap angle on the smaller pulley is at least 120° for a friction drive belt. For a toothed belt, the wrap angle can be closer to 90°. Any less than this can cause the belt to skip teeth.

Pulleys can be arranged in combinations that provide mechanical advantage along with the speed reduction. For any given pair of pulleys, it is usually best to not exceed a ratio of 8 to 1, and 6 to 1 is a

Components and Hardware   143

FIGURE 3.60   Positive drive (toothed) pulley and belt.

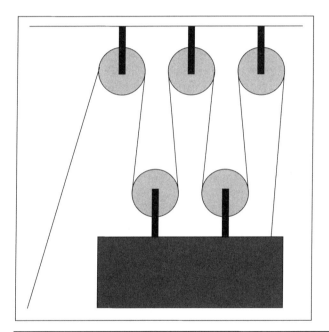

FIGURE 3.61   Compound drive.

reasonable maximum. If a greater ratio is required, it is best to use a compound drive of several pulleys. This is illustrated in Fig. 3.61.

3.7.5   Servomechanisms

A *servomechanism* is a combination of mechanical and control hardware that uses feedback to affect the control of a system. The feedback is in the form of an error or difference between the sensed parameter and

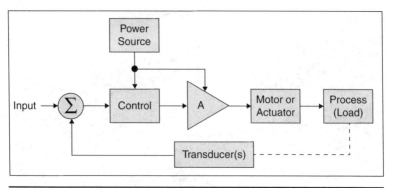

FIGURE 3.62 Servo system.

its desired value. Servomechanisms generally operate on the principle of negative feedback, where the error is subtracted from the output. A servomechanism is known as a closed-loop system. This has already been described in PID Control (section 2.2.2) and the Servomotor (section 3.6.4) sections of this book; however, it is important to mention that servomechanisms control not only position, speed, and torque, but may control variables such as temperature, pressure, or anything measurable. An example of a servomechanism that does not involve a servomotor is a hydraulic actuator that has its speed and position controlled by a spool valve using feedback from an externally mounted analog position sensor. Figure 3.62 illustrates the physical layout of a servo system.

3.7.6 Ball Screws and Belt-Driven Linear Actuators

A *ball screw* is a mechanical linear actuator that translates rotary motion to linear motion with little friction. A threaded shaft provides a spiral raceway for ball bearings, which act as a precision screw. The ball assembly acts as the nut, while the threaded shaft is the screw. They are made to close tolerances and are therefore suitable for use in situations where high precision is necessary.

The pitch of a ball screw determines the potential linear speed of a linear actuator as well as its ability to hold a load against the force of gravity. Higher pitch (more turns per inch) provides more precision and a greater ability to stop a vertical load from turning the screw, but it requires a higher motor speed to move at the same rate as a lower pitch screw. Figure 3.63 shows a 12 mm ball screw and nut with a 4 mm screw pitch.

Belt-driven linear actuators use a toothed belt and gears to move a carriage attached to the belt. The linear speed of a belt-driven actuator is typically faster than that of a ball screw but can be less rugged and precise. Belt-driven actuators are more likely to slip with a heavy load; toothed belts can even be damaged if the load is beyond the rating of the actuator as teeth can be stripped off the belt. Figure 3.64 shows a cutaway of an actuator showing the belt inside the sealed housing.

Components and Hardware 145

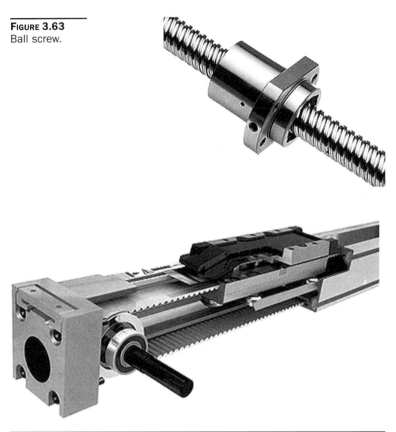

FIGURE 3.63
Ball screw.

FIGURE 3.64 Belt-driven actuator.

3.7.7 Linkages and Couplings

A *linkage* is a combination of rigid elements and hinges or joints that constrain element movement. A linkage can be used to multiply or translate force or motion between mechanisms or components. Examples of linkages are scissor lifts and four bar linkages.

The simplest linkage is the lever. By pinning a point on the length of the lever to a fixed location, the lever pivots around that point, the fulcrum. A point nearer to the fulcrum will rotate in a smaller arc than a point farther from the fulcrum, creating a multiplication of distance or velocity. With this comes a reduction in the force output; the reverse of this is also true, with a large movement driving a smaller one, the force output is multiplied proportionally. This is known as a mechanical advantage.

A four bar linkage is a group of four joints and four bars that allows points on the linkage to move in restrained ways when one or two of the joints are fixed in a location. Depending on the lengths of the bars and whether the joints can rotate through a full circular motion, different arcs and motions can be created, as shown in Fig. 3.65.

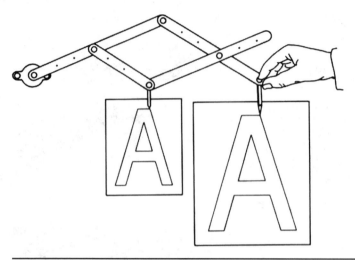

FIGURE 3.65   Four bar linkages.

This is a type of mechanism that uses a four bar linkage with two degrees of freedom known as a pantograph. It allows shapes to be duplicated at scaled sizes—a tool especially useful for engraved lettering. Similar arrangements can be used to constrain movement in mechanisms.

A common type of mechanism that uses a toggle linkage is the clamp. A toggle mechanism is a type of four bar linkage that folds and locks at a certain position. Some pneumatically driven clamps perform a rotate-and-lower movement that is partially cam based, but a manual toggle clamp uses a four bar linkage to produce a high clamping force, as shown in Fig. 3.66. A well-known manufacturer of toggle and other types of clamps is DE-STA-CO.

A *coupling* connects two shafts or rotating members together. A coupling can be rigid or flexible. A rigid coupling has the advantage of keeping shafts precisely aligned and holding the ends very securely. Flexible couplings allow for misalignment and tend to dampen vibration. Helical flexible couplings are commonly used to drive a shaft with a servomotor, AC motor, or DC motor in automated systems. Another name for a helical coupling is a beam coupling; an example is shown in Fig. 3.67.

A Lovejoy or spider coupling has two metal-toothed hubs and an elastomer insert or "spider" to reduce vibrations. The three parts fit together with a press fit. These are also used extensively in servo systems but are not as forgiving to angular misalignment as helical couplings.

Universal joints allow shafts to be driven at an angle, as do gear joints. These are often used in power transmission. For lighter applications, hoses or bellows-type couplings are sometimes used for nonaligned shafts.

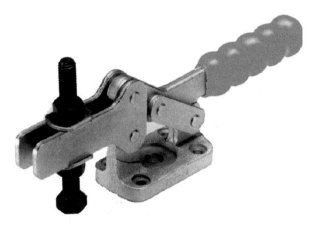

FIGURE 3.66 Manual toggle clamp.

FIGURE 3.67
Helical coupling.

3.7.8 Clutches and Brakes

A *clutch* allows two rotating elements to be engaged and disengaged with each other. The most common type of clutch is a friction clutch. There are several different types of friction clutch, including conical, radially expanding, contraction band, and friction disk clutches. It is common for the driving and driven members of friction clutches to be held in contact with each other by spring pressure. They may then be actuated to the open position by pneumatic, hydraulic, electrical, magnetic, or even centrifugal force. Clutch surfaces can be made of a variety of materials, including metals, ceramics, resin-wire fabric, and rubbers with a high durometer.

Brakes are used to rapidly stop a rotating member, also using frictional methods. Typically, brakes use a disk or pad to contact a rotating surface, such as a plate or shaft. In industrial applications, most brakes are composed of semimetallic, organic, or ceramic materials. Like clutches, brakes may be actuated in a number of ways, including pneumatics and hydraulics, electrical, or mechanical methods.

A *clutch brake* is a device that either engages a clutch to provide motion from a rotating to a nonrotating member or quickly brings the driven member to a stop when the clutch is not engaged. Clutch brakes are used extensively in conveyor applications and often in other servo and motor-driven machinery.

3.8 Structure and Framing

Machines are often built with a welded or bolted steel frame as a base structure. Today, however, more and more machines are built of aluminum extrusion or pipe-based structures. There are also many small structural assembly components available from machine part vendors.

Electrical components must be kept in an enclosed space to protect components or prevent personnel from contact with voltages. There are a number of classifications associated with this protection.

3.8.1 Steel Framing

Most steel frames are welded for rigidity and permanence. Steel tubing, flat-stock, and angle pieces are cut to length and welded together, generally using a jig, clamp, or fixture to ensure alignment. Because steel rusts or oxidizes, it usually has to be cleaned by grinding before being primed and painted. For food-processing or medical applications, frames are often made of stainless steel to eliminate the need for paint, which can cause contamination.

Premade welded steel bases can be purchased as a standard product from various manufacturers. These generally consist of a welded tube steel base, a ground steel top plate for mounting components, and adjustable feet and holes for bolting them to the floor (leveling feet). An example is shown in Fig. 3.68.

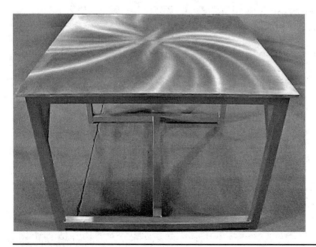

FIGURE 3.68 Welded machine base.

Frames may also be bolted together. This is often done for one of two reasons. One, the relative cost compared with welding is much cheaper; two, the ability to unbolt the frame later makes for easy transport or modification. Using fasteners to build a frame is not as desirable as welding, especially on heavy-duty frames that may be subject to vibration.

If components are to be mounted to a frame, a piece of steel is often welded to the frame to be machined to a specific thickness afterward. This *machine pad* can then be ground, drilled, and tapped for precise location of a component or subassembly.

During the welding and machining process, stresses are induced into the welded frame. Prior to assembly, these stresses are often relieved by heating the frame in an oven or attaching a *stress-relieving* vibrator to the frame. This is especially important on larger frames or frames that have critical dimensions.

Shims

A *shim* is a thin piece of material used to fill a space or increase a dimension by a slight amount. It is used as an adjustment method and is often placed between two objects that have been bolted together as a spacer.

Shim stock can be purchased in varying widths and thicknesses to be cut into desired sizes. For industrial machinery, shim stock is usually made of metal, although plastic composites are sometimes used. Shim stock is also available in a laminated foil, which can be peeled off a stack to build up a surface.

Dowels and Dowel Pins

A dowel or dowel pin is a solid cylindrical piece of material that can be pressed into a hole as a locating device. Dowels in industrial machinery are usually made of hardened steel. Dowel material is usually machined to a very tight tolerance. This material is available in long sections called dowel rods, which are then cut into pins. The ends of the pins are then usually lightly chamfered. Figure 3.69 shows a pair of dowel pins.

FIGURE 3.69
Dowel pins.

Dowel pins may have a slightly smaller diameter than the hole into which they are inserted so that they float freely, or they may have the same diameter as the hole, which is then reamed for a press fit.

Using bolts as fasteners introduces mechanical play into the positioning of the mounted piece. Typically, fasteners have clearances because of the oversizing of the through holes. This is proportional to the size of the fastener. By using dowels for precise location, this mechanical play can be reduced by up to ten times or eliminated altogether. The use of dowels does often increase the amount of assembly and disassembly time for components, but when precision is required, it is well worth the added cost.

Dowels are not meant to provide structural support. Any more than two dowels in a single assembly is not recommended. As soon as a side load is applied a few times, the brunt of the force will be transferred to one of the dowels, making the others redundant. It also makes it very difficult to reassemble the pieces if they are ever taken apart. Fasteners or keys and slots should be used to take the side load forces. Dowels should be used only as a precision locating method.

Keys, Keyseats, and Fixture Keys

A key is used to transmit torque between a shaft and hub. If both shaft and hub have a rectangular slot or keyseat placed in the axial direction, a key can be assembled into the grooves, providing a positive surface for the motor shaft to exert force against.

Keyseats or keyways are sized based on the diameter of the shaft in both width and depth. ANSI provides a table for sizing the key size and keyseat depth. Key stock and bar stock are available in standard or metric dimensions, and their cross sections may be rectangular or square-shaped.

Figure 3.70 shows a keyed motor shaft with a key inserted. Since this key is captured (that is, the keyway does not extend to

FIGURE 3.70
Keyed motor shaft.

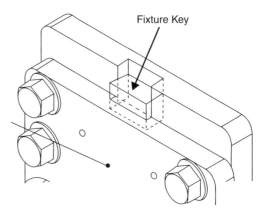

FIGURE 3.71
Fixture key.

the end of the shaft), it must be inserted before sliding it into a coupling.

Fixture keys are used to accurately locate fixtures or machine components. Typically, they are assembled into a hole or a machined slot. There are many different types of fixture keys, from simple square or rectangular keys inserted into slots in two mating surfaces (Fig. 3.71) to stepped or sine-type keys. They are made in metric or standard dimensions, usually of hardened steel. Socket head screws are sometimes threaded into the key for fastening.

Unlike dowels, fixture keys are designed to spread a load over a surface and are, therefore, appropriate for perpendicular support, that is, "side loading."

Machine Pads and Grind Spacers

Machine pads are pieces of flat stock that are welded to a frame in order to have something mounted to them. They often have tapped or through holes to accommodate the bolts for the mating assembly. There are two important reasons for using machine pads rather than mounting directly to a frame: to provide thickness for tapping or structural integrity and to provide a surface that can be ground to make it flat and parallel to other surfaces.

For large assemblies that are to be mounted onto a machine pad or other flat surface, grind spacers are often designed into a machine. These are pieces of metal that are purposely inserted between the flat machined surface on a frame and an assembly. This allows them to be ground to different thicknesses to make up for small differences in parallel surfaces. Rather than having to grind a large assembly or pads on a frame, the spacers can be easily removed and ground individually.

Fasteners

Fasteners are any device used to attach two pieces of material together. They include such classifications as rivets, bolts, and screws.

A rivet is a fastener used to attach items permanently. There are several different types of rivets including solid, tubular, or blind rivets—also known as "pop" rivets. Rivets consist of a shaft with a head on one end and are deformed after being inserted through a hole that passes through both pieces to be joined. This creates a connection that must be ground or drilled to remove. Solid rivets are typically deformed with a hammer, a rivet compression tool, or a crimping tool, which may be hydraulic, pneumatic, or electromagnetic. These types of riveting methods are used near the edge of the fastened materials so that the riveting tool can access both sides of the rivet.

For connections made away from the edge of the materials, a blind rivet, as shown in Fig. 3.72, is used. This is a tubular rivet with a large head that has a shaft with a mandrel through the center. The rivet is inserted through the hole and a special tool is used to pull the mandrel head into the rivet. The mandrel is designed to break off when enough force is applied.

Rivets may be made of steel, aluminum, or various other metals. In industrial automation, one of the most common uses of rivets is to attach wireway or other components permanently to an electrical backplane.

A *bolt* or *screw* is a fastener with a threaded shaft with a head on one end that is used to apply torque, driving the fastener into a threaded hole or a nut. Screws are usually used to drive into a hole without the use of a nut and often create their own threads in the hole, whereas bolts may be threaded into a prethreaded hole or into a nut. Bolt and screw heads come in many different shapes, depending on the type of tool that is used to drive them. Socket head and hex head are common forms of bolts, while Phillips or slotted heads are most common with screws. Screw heads can be pan or dome-shaped, round, countersunk, or several other forms. Special forms of bolt or screw heads also include hex socket, Robertson—or square drive—Torx, spanner head, and a variety of so-called "security" heads.

FIGURE 3.72
Blind or pop rivet.

Screws and bolts are usually threaded in a right-handed direction, meaning that the fastener must be turned clockwise to tighten and counterclockwise to loosen it. Sizes of bolts and screws generally fall into standard (SAE) and metric dimensions, which specify the shaft diameter and thread pitch.

3.8.2 Aluminum Extrusion

Profiles of extruded aluminum are frequently used for machine guarding but can also be used to build machines of substantial size. Aluminum sections are available in both metric and standard sizes from a variety of manufacturers. There is a wide range of available accessories, such as brackets, fasteners, hinges, plastic covers, and caps. Profiles are square or rectangular in their cross sections and include a "T-slot" in the side for fasteners, panels, or the routing of cables or hoses. Common vendors of these extrusion products include 80/20, Item and Bosch. Figure 3.73 illustrates hardware for joining two pieces of aluminum extrusion.

Aluminum extrusion also comes in a variety of different colors, although the natural silver/gray anodized color is most common.

While aluminum profile is more expensive than an equivalent length of tube steel, this is offset by the cost of welding, painting, and labor. Aluminum is not subject to rust and is usually anodized for hardness and electrical resistance.

3.8.3 Piping and Other Structural Systems

For lighter-duty applications, structures may be built of threaded pipe, angle, and flat stock. As with aluminum extrusion, these systems are sold by several different companies. Creform is probably

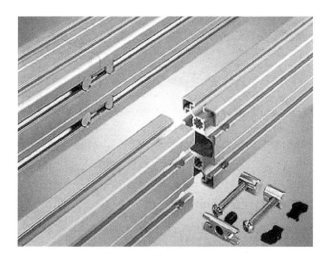

FIGURE 3.73 Aluminum extrusion. (*Courtesy of Bosch.*)

FIGURE 3.74 Pipe structure rack. (*Courtesy of Creform.*)

the best known of the threaded pipe systems. These types of structural systems are often used for carts, roller conveyor stands, and racks. Automation elements, such as pick-to-light sensors and gravity conveyor rollers, can also be mounted using accessories from these systems. Figure 3.74 shows a bin rack made of threaded pipe.

On a smaller scale, other vendors, such as Misumi, make structural items to mount sensors, gauges, or other devices. These items, sometimes known as "tinkertoys" (not to be confused with the wooden toy of the same name), are available in metric and standard sizes. Small brackets, bearings, gearing systems, and various other items are also available from this manufacturer.

3.8.4 Electrical Enclosures and Ratings

Control and electrical components are usually housed in metal or nonmetallic enclosures. Enclosures may be made of steel, galvanized sheet metal, fiberglass, or plastic. These enclosures are available in a wide range of sizes and configurations and from various

Components and Hardware 155

FIGURE 3.75 NEMA 12 electrical enclosure.

manufacturers. They typically have either screw covers or hinged doors for access to internal components and wiring.

Sizes range from small junction boxes or push-button boxes to large multidoor enclosures. Enclosures are rated for suitability for different environments by NEMA and IEC, as explained in the next section. A NEMA 12 enclosure is shown in Fig. 3.75. Larger enclosures are often manufactured with holes to mount a flange-mounted disconnect on the side by the door. These may be specified for disconnects from Allen-Bradley, Square D, Cutler-Hammer, or other major manufacturers.

Other accessories, such as filtered vents, fans, or even air conditioners, are available from enclosure manufacturers. Fluorescent lighting that is activated when a door is opened is another standard accessory.

Most enclosures are manufactured with studs to hold a metal backplane to mount components on. Backplanes may be steel or galvanized metal and are usually grounded during panel fabrication. Well-known electrical enclosure manufacturers include Hoffman and Rittal.

NEMA Ratings

Electrical enclosures are given a rating by NEMA. The following definitions are from NEMA Standards Publication 250-2003, "Enclosures for Electrical Equipment (100 Volts Maximum)":

In *nonhazardous locations*, the specific enclosure types, their applications, and the environmental conditions they are designed to protect against, when completely and properly installed, are as follows:

Type 1: Enclosures constructed for indoor use to provide a degree of protection to personnel against access to hazardous parts and to provide a degree of protection of the equipment inside the enclosure against ingress of solid foreign objects (falling dirt).

Type 2: Enclosures constructed for indoor use to provide a degree of protection to personnel against access to hazardous parts; to provide a degree of protection of the equipment inside the enclosure against ingress of solid foreign objects (falling dirt); and to provide a degree of protection with respect to harmful effects on the equipment due to the ingress of water (dripping and light splashing).

Type 3: Enclosures constructed for either indoor or outdoor use to provide a degree of protection to personnel against access to hazardous parts; to provide a degree of protection of the equipment inside the enclosure against ingress of solid foreign objects (falling dirt and windblown dust); to provide a degree of protection with respect to harmful effects on the equipment due to the ingress of water (rain, sleet, snow); and that will be undamaged by the external formation of ice on the enclosure.

Type 3R: Enclosures constructed for either indoor or outdoor use to provide a degree of protection to personnel against access to hazardous parts; to provide a degree of protection of the equipment inside the enclosure against ingress of solid foreign objects (falling dirt); to provide a degree of protection with respect to harmful effects on the equipment due to the ingress of water (rain, sleet, snow); and that will be undamaged by the external formation of ice on the enclosure.

Type 3S: Enclosures constructed for either indoor or outdoor use to provide a degree of protection to personnel against access to hazardous parts; to provide a degree of protection of the equipment inside the enclosure against ingress of solid foreign objects (falling dirt and windblown dust); to provide a degree of protection with respect to harmful effects on the equipment due to the ingress of water (rain, sleet, snow); and for which the external mechanism(s) remain operable when ice laden.

Type 3X: Enclosures constructed for either indoor or outdoor use to provide a degree of protection to personnel against access to hazardous parts; to provide a degree of protection of the equipment inside the enclosure against ingress of solid foreign objects (falling dirt and windblown dust); to provide a degree of protection with respect to harmful effects on the equipment due to the ingress of water (rain, sleet, snow); that provides an additional level of protection against corrosion and that will be undamaged by the external formation of ice on the enclosure.

Type 3RX: Enclosures constructed for either indoor or outdoor use to provide a degree of protection to personnel against access to hazardous parts; to provide a degree of protection of the equipment inside the enclosure against ingress of solid foreign objects (falling dirt); to provide a degree of protection with respect to harmful effects on the equipment due to the ingress of water (rain, sleet, snow); that will be undamaged by the external formation of ice on the enclosure that provides an additional level of protection against corrosion; and that will be undamaged by the external formation of ice on the enclosure.

Type 3SX: Enclosures constructed for either indoor or outdoor use to provide a degree of protection to personnel against access to hazardous parts; to provide a degree of protection of the equipment inside the enclosure against ingress of solid foreign objects (falling dirt and windblown dust); to provide a degree of protection with respect to harmful effects on the equipment due to the ingress of water (rain, sleet, snow); that provides an additional level of protection against corrosion; and for which the external mechanism(s) remain operable when ice laden.

Type 4: Enclosures constructed for either indoor or outdoor use to provide a degree of protection to personnel against access to hazardous parts; to provide a degree of protection of the equipment inside the enclosure against ingress of solid foreign objects (falling dirt and windblown dust); to provide a degree of protection with respect to harmful effects on the equipment due to the ingress of water (rain, sleet, snow, splashing water, and hose directed water); and that will be undamaged by the external formation of ice on the enclosure.

Type 4X: Enclosures constructed for either indoor or outdoor use to provide a degree of protection to personnel against access to hazardous parts; to provide a degree of protection of the equipment inside the enclosure against ingress of solid foreign objects (windblown dust); to provide a degree of protection with respect to harmful effects on the equipment due to the ingress of water (rain, sleet, snow, splashing water, and hose directed water); that provides an additional level of protection against corrosion; and that will be undamaged by the external formation of ice on the enclosure.

Type 5: Enclosures constructed for indoor use to provide a degree of protection to personnel against access to hazardous parts; to provide a degree of protection of the equipment inside the enclosure against ingress of solid foreign objects (falling dirt and settling airborne dust, lint, fibers, and flyings); and to provide a degree of protection with respect to harmful effects on the equipment due to the ingress of water (dripping and light splashing).

Type 6: Enclosures constructed for either indoor or outdoor use to provide a degree of protection to personnel against access to hazardous parts; to provide a degree of protection of the equipment inside the enclosure against ingress of solid foreign objects (falling dirt); to provide a degree of protection with respect to harmful effects on the equipment due to the ingress of water (hose directed water and the entry of water during occasional temporary submersion at a limited depth); and that will be undamaged by the external formation of ice on the enclosure.

Type 6P: Enclosures constructed for either indoor or outdoor use to provide a degree of protection to personnel against access to hazardous parts; to provide a degree of protection of the equipment inside the enclosure against ingress of solid foreign objects (falling dirt); to provide a degree of protection with respect to harmful effects on the equipment due to the ingress of water (hose directed water and the entry of water during prolonged submersion at a limited depth); that provides an additional level of protection against corrosion and that will be undamaged by the external formation of ice on the enclosure.

Type 12: Enclosures constructed (without knockouts) for indoor use to provide a degree of protection to personnel against access to hazardous parts; to provide a degree of protection of the equipment inside the enclosure against ingress of solid foreign objects (falling dirt and circulating dust, lint, fibers, and flyings); and to provide a degree of protection with respect to harmful effects on the equipment due to the ingress of water (dripping and light splashing).

Type 12K: Enclosures constructed (with knockouts) for indoor use to provide a degree of protection to personnel against access to hazardous parts; to provide a degree of protection of the equipment inside the enclosure against ingress of solid foreign objects (falling dirt and circulating dust, lint, fibers, and flyings); and to provide a degree of protection with respect to harmful effects on the equipment due to the ingress of water (dripping and light splashing).

Type 13: Enclosures constructed for indoor use to provide a degree of protection to personnel against access to hazardous parts; to provide a degree of protection of the equipment inside the enclosure against ingress of solid foreign objects (falling dirt and circulating dust, lint, fibers, and flyings); to provide a degree of protection with respect to harmful effects on the equipment due to the ingress of water (dripping and light splashing); and to provide a degree of protection against the spraying, splashing, and seepage of oil and noncorrosive coolants.

In *hazardous locations*, when completely and properly installed and maintained, Type 7 and 10 enclosures are designed to contain an internal explosion without causing an external hazard. Type 8 enclosures are designed to prevent combustion through the use of oil-immersed equipment. Type 9 enclosures are designed to prevent the ignition of combustible dust.

Type 7: Enclosures constructed for indoor use in hazardous (classified) locations classified as Class I, Division 1, Groups A, B, C, or D as defined in NFPA 70.

Type 8: Enclosures constructed for either indoor or outdoor use in hazardous (classified) locations classified as Class I, Division 1, Groups A, B, C, and D as defined in NFPA 70.

Type 9: Enclosures constructed for indoor use in hazardous (classified) locations classified as Class II, Division 1, Groups E, F, or G as defined in NFPA 70.

Type 10: Enclosures constructed to meet the requirements of the Mine Safety and Health Administration, 30 CFR, Part 18.

Tables from NEMA 250-2003 are contained in App. D.

IEC and IP Ratings

The IEC also rates both enclosures and other electrical devices for suitability in various environments. A rating known as IP, for Ingress Protection, is used. This can be converted to an equivalent NEMA rating using Table 3.2.

The IP Code consists of the letters *I* and *P* followed by two digits or one digit and one letter and an optional letter. As defined in

IP Code	Minimum NEMA Enclosure Rating to Satisfy IP Code
IP20	1
IP54	3
IP65	4, 4X
IP67	6
IP68	6P

TABLE 3.2 IP/NEMA Equivalents

international standard IEC 60529, IP Code classifies and rates the degrees of protection provided against the intrusion of solid objects (including body parts like hands and fingers), dust, accidental contact, and water in mechanical casings and with electrical enclosures.

The standard aims to provide users with more detailed information than vague marketing terms such as "waterproof." However, no edition of the standard is openly published for unlicensed readers, hence leaving room for varying interpretation.

The digits (IP numerals) indicate conformity with the conditions summarized in the tables below. Where there is no protection rating with regard to one of the criteria, the digit is replaced with the letter X.

For example, an electrical socket rated IP22 is protected against insertion of fingers and will not be damaged or become unsafe during a specified test in which it is exposed to dripping water. IP22 or 2X are typical minimum requirements for the design of electrical accessories for indoor use.

Solids, First Digit

The first digit indicates the level of protection that the enclosure provides against access to hazardous parts (for example, electrical conductors, moving parts) and the ingress of solid foreign objects, as shown in Table 3.3.

Level	Object Size Protected Against	Effective Against
0	—	No protection against contact and ingress of objects
1	> 50 mm	Any large surface of the body, such as the back of a hand, but no protection against deliberate contact with a body part
2	> 12.5 mm	Fingers or similar objects
3	> 2.5 mm	Tools, thick wires, etc.
4	> 1 mm	Most wires, screws, etc.
5	Dust protected	Ingress of dust is not entirely prevented, but it must not enter in sufficient quantity to interfere with the satisfactory operation of the equipment; complete protection against contact
6	Dust tight	No ingress of dust; complete protection against contact

Table 3.3 IP Code for Solids (First Digit)

Liquids, Second Digit

The second digit categorizes the protection of equipment inside the enclosure against harmful ingress of water, as shown in Table 3.4.

Level	Protected Against	Testing For	Details
0	Not protected	—	—
1	Dripping water	Dripping water (vertically falling drops) shall have no harmful effect.	Test duration: 10 minutes. Water equivalent to 1 mm rainfall per minute
2	Dripping water when tilted up to 15°	Vertically dripping water shall have no harmful effect when the enclosure is tilted at an angle up to 15° from its normal position.	Test duration: 10 minutes. Water equivalent to 3 mm rainfall per minute
3	Spraying water	Water falling as a spray at any angle up to 60° from the vertical shall have no harmful effect.	Test duration: 5 minutes. Water volume: 0.7 L per minute Pressure: 80 to 100 kN/m²
4	Splashing water	Water splashing against the enclosure from any direction shall have no harmful effect.	Test duration: 5 minutes. Water volume: 10 L per minute Pressure: 80 to 100 kN/m²
5	Water jets	Water projected by a nozzle (6.3 mm) against enclosure from any direction shall have no harmful effects.	Test duration: at least 3 minutes. Water volume: 12.5 L per minute Pressure: 30 kN/m² at distance of 3 m
6	Powerful water jets	Water projected in powerful jets (12.5 mm nozzle) against the enclosure from any direction shall have no harmful effects.	Test duration: at least 3 minutes. Water volume: 100 L per minute Pressure: 100 kN/m² at distance of 3 m

TABLE 3.4 IP Code for Liquids (Second Digit) (*Continued*)

Level	Protected Against	Testing For	Details
7	Immersion up to 1 m	Ingress of water in harmful quantity shall not be possible when the enclosure is immersed in water under defined conditions of pressure and time (up to 1 m of submersion).	Test duration: 30 minutes Immersion at depth of 1 m
8	Immersion beyond 1 m	The equipment is suitable for continuous immersion in water under conditions that shall be specified by the manufacturer. Normally, this will mean that the equipment is hermetically sealed. However, with certain types of equipment, it can mean that water can enter but only in such a manner that it produces no harmful effects.	Test duration: continuous immersion in water Depth specified by manufacturer

TABLE 3.4   IP Code for Liquids (Second Digit) (Continued)

Level	Protected against Access to Hazardous Parts With
A	Back of hand
B	Finger
C	Tool
D	Wire

TABLE 3.5   IP Codes for Access to Hazardous Parts

Additional Letters

The standard defines additional letters that can be appended to classify only the level of protection against access to hazardous parts by persons, as shown in Table 3.5.

Additional letters can be appended to provide further information related to the protection of the device, as illustrated in Table 3.6.

Mechanical Impact Resistance

An additional number has sometimes been used to specify the resistance of equipment to mechanical impact. This mechanical impact

Letter	Meaning
H	High-voltage device
M	Device moving during water test
S	Device standing still during water test
W	Weather conditions

TABLE 3.6 IP Codes Describing Test Conditions

Dropped IP Level	Impact Energy	Equivalent Drop Mass and Height
0	—	—
1	0.225 J	150 g dropped from 15 cm
2	0.375 J	250 g dropped from 15 cm
3	0.5 J	250 g dropped from 20 cm
5	2 J	500 g dropped from 40 cm
7	6 J	1.5 kg dropped from 40 cm
9	20 J	5.0 kg dropped from 40 cm
IK Number	**Impact Energy (joules)**	**Equivalent Impact**
00	Unprotected	No test
01	0.15	Drop of 200 g object from 7.5 cm height
02	0.2	Drop of 200 g object from 10 cm height
03	0.35	Drop of 200 g object from 17.5 cm height
04	0.5	Drop of 200 g object from 25 cm height
05	0.7	Drop of 200 g object from 35 cm height
06	1	Drop of 500 g object from 20 cm height
07	2	Drop of 500 g object from 40 cm height
08	5	Drop of 1.7 kg object from 29.5 cm height
09	10	Drop of 5 kg object from 20 cm height
10	20	Drop of 5 kg object from 40 cm height

TABLE 3.7 IK Impact Resistance Codes

is identified by the energy needed to qualify a specified resistance level, which is measured in joules (J). The separate IK number specified in EN 62262 has now superseded this measurement.

Although dropped from the third edition of IEC 60529 onward—and not present in the EN version—older enclosure specifications will sometimes be seen with an optional third IP digit denoting impact resistance. Newer products are likely to be given an IK rating instead. There is not an exact correspondence of values between the old and new standards. These codes are shown in Table 3.7.

CHAPTER 4

Machine Systems

The elementary actuators and techniques described in Chap. 3 are used in concert with one another to accomplish different tasks. For instance, motors, gearboxes, bearings, and belts are combined within a frame to form a conveyor, or pneumatic actuators; vibratory thrusters and sensors are built into a vibratory part feeder. Manufacturers often concentrate their expertise into combining these techniques into standard products, whereas custom machine builders use these systems to create unique combinations for each application.

4.1 Conveyors

Conveyors are used to move objects or substances from one point to another. They can take many forms and are usually driven by a motor, air, or gravity. Large conveyor systems often have a centralized control system controlled by a PLC. Because of the long distances associated with conveyor systems, sensors and actuators were historically often operated at 120VAC; however, with advances in technology using distributed I/O and modern safety regulations, 24VDC systems are now common. Figure 4.1 shows a conveyor system in a cotton-testing facility.

Motors on these large systems are usually of the three-phase 480VAC variety. This requires I/O and motor power to be run separately if 24VDC I/O is used because of the potential of electrical interference. Distributed I/O using communication methods such as Profibus, Ethernet, or DeviceNet requires additional cabling that is also usually attached to the frame of the conveyors. A local disconnect is often provided near each motor and may be monitored by the control system. Safety devices such as E-Stop push buttons and cable-pull actuated E-Stops are also generally mounted to the frames.

HMIs often depict the layout of the system, showing the status of the system components along with production or packaging machines integrated with the system. Conveyor control systems can be very elaborate and have hundreds or thousands of I/O points. They also often use multiple variations of the types of conveyors described in this chapter.

Chapter Four

FIGURE 4.1 Conveyor system.

4.1.1 Belt Conveyors

A belt conveyor consists of two or more pulleys or rollers with a continuous loop of material with the conveyor belt rotating around them. One or both of the pulleys may be powered, moving the belt and the material on the belt forward. Powered pulleys or rollers are called drive or driven elements, while the undriven rollers or pulleys are called idlers. Idlers may also be located on the underside of the conveyor for support of the return strand of belt. Motors for belt conveyors are typically located at the head, or pulling end, of the conveyor. For reversing conveyors, the motor may be located in the middle.

Belts may be made of many different materials ranging from rubber or plastic compounds to metal mesh. Many belts are made of composites with an under layer for strength and a cover material to protect the product. Belt conveyors are usually used in applications requiring a solid surface, where materials cannot be easily passed across rollers. Belt materials are often chosen based on strength requirements or load, the amount of friction required, and the environment that they are exposed to. Cleats and sidewalls may be attached to the belt surface to help confine materials or reduce the need for high-friction surfaces that may damage products. Cleat spacing and durability are key factors in the choice of material and bonding methods for a cleated belt.

Machine Systems

FIGURE 4.2 Cleated incline belt conveyor. (*Courtesy of Nalle Automation Systems.*)

If belt conveyors are used on an incline or decline, the friction coefficient of the belt is typically high. A nose-over section is often placed on the top, bottom, or both sections of incline belt conveyors to allow for easy transitions of material from other conveyors or into hoppers. Figure 4.2 shows a small cleated conveyor on wheels used for packaging.

Belt conveyors are one of the least expensive types of conveyor. They typically have a metal framework with rollers at each end. The belt may be pulled across a flat surface or bed. For heavier loads, it may also move across additional rollers. These are known as slider bed and roller bed conveyors, respectively. To ensure the belt is at the proper tightness and tracks well across the rollers, the end roller is often adjustable. Rollers may also be crowned to ensure belt centering.

4.1.2 Roller Conveyors

Roller conveyors can take several forms: they may be powered or unpowered, belt or chain driven, or even series of individually powered rollers.

Rollers are usually a metal shell with a shaft on each end. Depending on the weight and material being conveyed, rollers may be thin-wall aluminum or heavier-gauge steel, rubber coated or individual gravity "skate wheels." Thin-wall rollers are easily bent, dented, or cut and are not suitable for all applications, but they are often used for package handling. Axles on these rollers are often spring-loaded for easy removal.

Roller conveyors are usually used for moving packages with flat bottoms, like boxes. Rollers should be spaced so that at least three rollers are underneath the package at any time. Rollers may be driven using various methods. A line shaft may be placed along the length of the conveyor with individual urethane belts attached to each roller from spools on the shaft. Another method of driving rollers is to place either a flat or V belt on the underside of the rollers.

Metal chain can also be used to drive the rollers. A single chain can be used to drive all of the rollers or rollers can be linked together with roller-to-roller links. A greater number of sprocket teeth in contact with the chain allows for heavier loading.

Roller conveyors present special challenges when used on a curved section. Rollers must be spaced farther apart at the outside edge of the curve. Using a double section of rollers with more rollers on the outside section can mitigate this. Some rollers are even made that are larger on one end than the other. One note on curved conveyors—product should never be accumulated on a curve.

Figure 4.3 illustrates part of a roller conveyor system for cardboard containers. This section is known as a "merge."

An interesting product sometimes used for roller conveyors is the individually powered roller. These are essentially cylindrical motors with fixed shafts. These are usually DC powered and can be used to drive products section by section.

Gravity roller conveyors may be of the roller, or "skate wheel," type. These unpowered conveyors are usually used in short horizontal runs where operators push products from one end to the other or when products drop from one level to another. Skate wheel conveyors are often placed on a wheeled frame so that it can be moved from one location to another. Another closely related nonpowered conveying

FIGURE 4.3 Roller conveyor.

device is the ball table, which allows products to be moved in any direction by pushing them across a table embedded with large ball bearings. These are often used when moving totes of parts in machine loading and unloading areas.

4.1.3 Chain and Mat Conveyors

Chain conveyors use a continuous chain that runs from one sprocket to another at each end of a frame. Pendants or containers may be attached to the chain for product containment and transport. The most common type of chain conveyor is the tabletop chain conveyor, which has flat plates connected to the chain. Cleats are sometimes added to these plates for product separation and indexing.

Chain conveyors often use parallel strands of chain mounted to dual sprockets or gears on each end of an axle or shaft. This allows for devices such as lifts, stops, or transfers to be mounted between the chains. Tabletop chains with slats or plates can then be used to move pallets or products between these devices.

Tabletop chains may be composed of a thermoplastic material or metal. The chain is usually contained in a channel between the sprockets that guide it. The plates can also be made in such a way that they overlap and can turn, allowing for curved conveyor runs. Another term for this type of conveyor is *multiflexing* since the chain is flexible both sideways and on an incline. An example of a tabletop chain conveyor is shown in Fig. 4.4. Because the plates have space between them, they are also effective for drainage or airflow—an important consideration if working with metal machine parts.

Tabletop chain conveyors are not placed under tension like a belt conveyor since a sprocket is used to drive it. A catenary is typically used at the ends on the chain; this is a hanging loop that allows for an easy return of the chain on the underside of the conveyor frame. The catenary runs from the sprocket over a "shoe" and into containment by Teflon guides underneath.

Chain conveyors may also be used for suspending parts or pendants. A common application for this type of conveyor is for paint booths or ovens. In this case, the chain is almost always metal with hooks located at intervals along the chain.

The mat top conveyer is closely related to the single column of links used in a chain conveyor. This type of conveyor uses multiple columns of links chained together in a mat. Although not as flexible as tabletop chains on curves, mat top conveyors can generally support more weight.

Chain and mat conveyors are usually driven by AC motors, often with variable speed drives for speed control. Chain and mat conveyors with cleats may also be driven with a servo for indexing purposes. Typically this is done using a sensor at the cleat for stopping the indexing motion and verifying position.

Chapter Four

FIGURE 4.4 Tabletop chain conveyor.

4.1.4 Vibrating Conveyors

Vibrating conveyors are used for moving bulk materials. Sometimes called shakers or shaker tables, they have a solid conveying surface with sides to contain the material being conveyed.

Vibrating conveyors operate on the natural frequency principle. With only a small energy input, an object can be made to vibrate at some frequency by alternately storing and releasing energy using supporting springs. The drive mechanism is usually an electric motor with a fixed eccentric shaft or rotating weight. A flat pan vibrating conveyor will convey most materials at a 5° incline from horizontal.

Food grade applications use vibrating conveyors extensively. Because vibrating conveyors are often made of stainless steel and can be easily coated with materials such as Teflon, they are suitable for wash-down and corrosive environments. They are low maintenance and excellent for sanitary applications. They are also used in applications for sorting, screening, classifying, and orienting parts.

Accessories for vibrating conveyors include counterbalance members for reducing reactions by generating an out-of-phase response to conveyor motion and weighted bases with isolation springs to reduce transmitted vibrations.

Air knife separators are an air-driven method of separating different weight materials. They are sometimes used with vibrating conveyors as a sanitary noncontact method of diverting material.

4.1.5 Pneumatic Conveyors

Pneumatic conveyors use pipes or ducts to transport materials using a stream of air. The most commonly transported materials using this method are dry pulverized or free-flowing powdery materials.

Carriers can also be transported using air. Items can simply be pushed from one location to another using a push or pull pressure system.

Following are three basic systems that are used to generate high-velocity airstreams for conveying:

1. Suction or vacuum systems use a vacuum created in the pipeline to draw the material with the surrounding air. The system is operated at a low pressure, usually 0.4 to 0.5 atm of pressure. This method is used mainly in conveying light free-flowing materials.

2. Pressure type systems use a positive pressure to push material from one point to the next. The system is ideal for conveying material from one loading point to a number of unloading points. It operates at a pressure of 6 atm and upward.

3. Combination systems use a suction system to convey material from a number of loading points and a pressure system to deliver it to a number of unloading points.

Air pressure may be generated using an industrial blower or fan. Alternatively, compressed air is sometimes used for small-volume applications.

4.1.6 Accessories

In addition to conveying components, there are various devices that are used to guide product in conveying systems.

Diverters are used to move product in a transverse direction to the direction of the conveyor. Sometimes called plows when used for bulk materials, diverters usually have a pivot point at one end. Diverters may be used to move product off the side of a conveyor onto spurs or guide product into lanes for sorting.

When guiding objects, it is important to consider the angle at which the diverter will operate. To ensure a smooth transition for items such as packages, it is best to operate the diverter at 30° to the conveyor flow. Under no circumstance can they be at greater than 45°. Bulk materials can generally operate at greater angles.

Diverters can be used with belt, roller, and chain conveyors. Pneumatic system diverters are also common. Diverters usually use air cylinders for actuation, but servo-operated diverters are also common where multiple positions are required.

Pushers are used to move objects at right angles to the conveyor. They are usually pneumatically operated and are often used in roller conveyors; however, they are not appropriate for belt conveyors.

Gates and lifts can be used to allow passage of personnel and vehicles. These are essentially self-contained conveyors on hinges.

Elevators are also commonly used to move product from one conveyor level to another. These may be pneumatically or motor operated and usually include a short length of conveyor within the lifting platform. With safety devices and traffic control sensors, these are often self-contained machines.

4.2 Indexers and Synchronous Machines

Indexers move objects a fixed distance for repetitive positioning and to avoid cumulative errors. They are often used to move objects being worked between fixed location stations. Walking beams and pick-and-place mechanisms also move objects from one location to another.

4.2.1 Rotary Cam Indexers

Rotary indexers are used to move actuators to fixed points in a circular path. They are built to move to discrete points and are typically available in 2- to 12-point configurations. Because they are cam driven, they can be moved at a high rate of speed and handle heavy loads. They can be driven by constant speed motors and can drive auxiliary actuators to perform other repetitive tasks as part of their operation. A common name for a rotary indexer with a machined platform for stations on top is a *dial table*. An example of a four-station dial table is shown in Fig. 4.5. Well-known manufacturers of these indexing devices include Camco and Stelron.

Sensors are used to detect when the indexer is within the "dwell" part of the index so station devices can operate on the product. A common device that is often used on dial tables is an overload clutch. If the motor tries to index the dial, but something is in the way, the dial will "break away" from the drive unit. A sensor is used to detect this condition, and the dial must be put back into position manually.

4.2.2 Synchronous Chassis Pallet Indexers

Synchronous chassis make use of a motor and line shaft to index pallets and synchronize devices performing operations around the chassis. Cams on the line shaft are used to operate devices in time with the movement of pallets, as well as control pallet movement

FIGURE 4.5 Dial table.

and dwell times. Synchronous chassis are more robust than systems using sensors and independent control of stations but less flexible and more difficult to reconfigure. Indexing drives and cam motions are specific to the machine application. Machine timing is mechanically fixed, so there is no risk of losing timing or position on individual workstations.

A clutch may be used to disconnect the drive mechanism from the chassis. This must be done while the cam is in the dwell, or nondriving part, of the cam motion. Otherwise, a chassis must be slowed at a gradual rate to reduce stress on the cam-driven mechanisms.

4.2.3 Walking Beams

A walking beam uses an X-axis and Z-axis configuration to repetitively index parts a fixed distance in a single direction. The X-axis moves forward with the Z-axis raised, carrying or pushing the part in the desired direction. The Z-axis is then lowered and the X-axis returned to its origin to begin another index. Walking beams are common in packaging and assembly operations because of their relatively low cost and repetitive accuracy. Axes may be pneumatically or servo actuated. Figure 4.6 illustrates the principle of a two-axis walking beam.

A variation of a walking beam for boxes, totes or flat products that does not require a Z-axis is a spring-finger walking beam. This is a horizontal axis with sloped fingers that are held in the raised position with springs. As the beam is moved backward underneath the product, the product pushes down the fingers; when the beam

Chapter Four

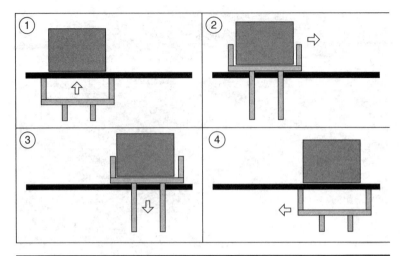

FIGURE 4.6 Walking beam.

moves in the forward direction, the fingers pop back up and propel the product forward.

4.2.4 Pick-and-Place

A pick-and-place is so named because it typically picks up an object and places it in another location. It consists of an X-axis, or horizontal axis; a Z-axis, or vertical axis; and a picking mechanism such as a mechanical gripper, vacuum cups, or even a magnet. If another

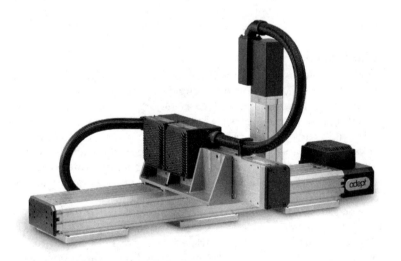

FIGURE 4.7 Three-axis robot used for pick-and-place. (*Courtesy of Adept.*)

horizontal axis, or Y-axis, is added, it is described as a *gantry*, which is further described in the robotic section 4.4 of this book.

Pick-and-place mechanisms may be pneumatically or servo driven, depending on speed requirements and the number of locations that must be accessed. It is common to see mechanisms that are a combination of both. Servo axes provide flexibility since their positions and speeds can be reprogrammed for different products. Figure 4.7 is a three-axis robot from Adept Robotics. A gripper can be attached to the lower end of the vertical axis, or Z-axis, creating a pick-and-place.

Other variations of the pick-and-place can be fabricated using linkages, cams, and other basic mechanisms. An example is a literature or sheet feeder that strips sheets of paper from a hopper or rack using vacuum cups and places them onto a flat surface.

4.3 Part Feeders

Part feeders supply components to a variety of manufacturing processes. They often serve as a buffer and part orientation device.

4.3.1 Vibratory Bowls and Feeders

Vibratory bowls and feeders use a variable amplitude controller to control a drive unit with spring thrusters oriented in the direction of part movement. Similar to the method of driving vibratory conveyors, leaf springs are mounted on a base unit oriented in the direction of the desired travel of the part. A bowl with special tooling and tracks sized to the component is then mounted to the other end of the springs to guide the parts. The tooling is also used to orient components within the bowl, guiding improperly oriented parts back into the bowl and allowing parts with the correct orientation to proceed. Figure 4.8 shows a vibratory bowl used to feed plastic caps.

Linear tracks also use vibratory drive units to move components in a straight line. Sensors and stops or gates may then be used to control the flow of parts along the tracks.

Drive units are available in electromagnetic and pneumatic drives. Parts are forced up a circular inclined track inside the bowl. The track length, width, and depth are carefully chosen to suit the application and component shape and size. Special track coatings are applied according to shape, size, and material of the component, which aids traction, minimizes damage to the product, and lowers acoustic levels.

Different materials travel better with different vibration frequencies. The amplitude and frequency of the electronic drive unit is generally set based on the optimum movement rate for the part being moved. Weights can also be added or removed from motors to adjust coarse feed rates. Spring constant values and lengths are other important considerations that affect part movement.

Chapter Four

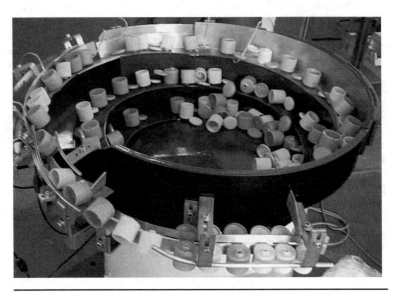

FIGURE 4.8 Vibratory bowl.

Vibratory feeders are used by most industries, including pharmaceutical, automotive, electronic, food, packaging, and metalworking. They are most commonly used in the assembly process as they align components for access by other mechanisms.

Vibratory hoppers and trays are also used to move bulk materials and are not always associated with orienting or singulating parts. These are more commonly used in the material handling and process industries for flow control.

4.3.2 Step and Rotary Feeders

Step feeders remove component parts from a hopper by either elevating parts with a single moving stepper plate onto stationary ledges, as illustrated in Fig. 4.9, or by counterrotating two stepper plates. Plates are moved on linear guides pushing product up out of the bin or hopper. Components are elevated until they reach the desired transfer height, generally feeding into a linear track or conveyor. Step feeders are often used on parts that are cylindrical and are not appropriate for vibrating feeders because of potential product abrasion.

Key features of a step feeder are that it operates quietly and without vibration. Width and thickness of the stepper plates are important variables to consider in feeder design. When considering using a step feeder over a vibratory system, it is important to remember that the part will be elevated considerably above the level of the hopper.

Centrifugal feeders, also referred as rotary feeders, have a conical central-driven rotor surrounded by a circular bowl wall. The feeder

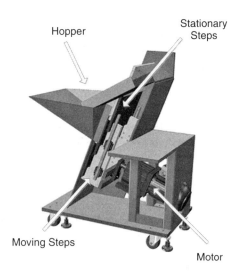

FIGURE 4.9
Step feeder.

separates component parts using rotary force; then parts revolve with high speed and are pulled to the outside of the bowl. As parts accumulate at the outside edge of the bowl, they tend to line up allowing for orientation and singulation of the parts.

Centrifugal feeders can be operated at higher feed rates than vibratory bowls. They are also better at handling oil-covered parts but do not work as well if parts tangle easily. Specialized tooling for parts orientation is used in much the same way as a vibratory bowl, with machined protrusions on the outside track of the feeder.

4.3.3 Escapements and Parts Handling

An *escapement* is a pair of actuators that allows for singulation of parts. If components are separated by a space, gates or pop-up stops can be used between parts as they move through the system. As a part is detected at the forward stop, another stop is raised behind the part. After the forward stop allows the part to exit and is raised again, the rear stop is lowered, allowing the part to move forward, after which the cycle is repeated.

Escapements may take the form of stops operating from above, below, or the sides. They may also use pressure against the side of the part itself; this is useful if parts do not have space between them.

Escapements are used in conveying systems, feeders, pallet indexing systems, and assembly. They are one of the basic elements of parts handling.

In general, parts handling involves controlling the movement of components within a system. A good rule of thumb is once you have control of randomized parts through orientation and singulation, you should never relinquish it. This means that if parts have been individualized, do not let them recombine. If they have been oriented, do not allow them to return to a mixed state.

Most part-handling techniques have been discussed in other sections of this chapter. Pushers and diverters, pick-and-places, and other actuator-based mechanisms are all examples of parts handling. Additional methods include using parallel urethane cord conveyors at an angle on the sides of parts to lift them out of moving conveyor pockets, using rubber wheels to accelerate low-friction parts for singulation, and using air to blow rejected or disoriented parts off a conveyor or out of a track. Actuators, air, and sensors can be used in many creative ways to create the desired part movement.

4.4 Robots and Robotics

A robot is an electromechanical machine that can perform tasks on its own or with guidance. Industrial robots are used widely throughout the manufacturing sector, and the various categories of these robots come in different configurations and sizes. Robots are most often driven by coordinated servo gear motors moving directly on axes; however, hydraulic robots are also used in some applications.

An *industrial robot* is defined by ISO 8373 as an "automatically controlled, reprogrammable, multipurpose manipulator programmable in three or more axes." In industry, the term *robotics* can be defined as the design and use of robot systems for manufacturing.

The most commonly used robot configurations are articulated robots, SCARA robots, and Cartesian coordinate robots (also known as gantry robots or x-y-z robots). Speed requirements, the positions that must be attained, and the cost of the system are factors that determine which type of configuration is generally used for a particular function.

4.4.1 Articulated Robots

An articulated robot is one that uses rotary joints to access its work space. Usually the joints are arranged in a "chain," so that one joint supports another farther in the chain. Another term for an articulated robot is a "robotic arm."

Articulated robots usually have anywhere from three to six joints. More than six joints are possible, but these robots generally fall into the custom category. Another term for this is "degrees of freedom," defined as the number of independent motions that make up the robot's area of operation. Joints are usually defined as J1-Jx, where x is the number of joints in the robot. J1 is the joint nearest the base of the robot, and other joints increment from there. Typically, J1 rotates horizontally around the robot base. Because of the cables that need to make their way through the various joints for servo power and position, joint rotation for J1 is usually less than 360°. Figure 4.10 shows a six-axis Denso robotic arm mounted on a base.

J2 and J3 usually operate in the vertical plane. Along with the rotation of J1, this allows the other joints to be placed close to nearly

Machine Systems 179

FIGURE 4.10 Robotic arm.

every point within the robot's operating envelope. J4, J5, and J6 typically act as manipulators, with the last joint, J6, usually being a rotary to which grippers or other devices are attached.

4.4.2 SCARA Robots

The SCARA acronym stands for Selective Compliant Assembly Robot Arm or Selective Compliant Articulated Robot Arm. These are usually of the four-axis variety, with J1 and J2 being horizontal rotary joints to access X-Y points, J3 being a Z-axis, and J4 being a rotary or T-axis mounted at the end of J3.

Because of the parallel axes of J1 and J2, the end of the vertical axis J3 is rigidly controlled in the X-Y position, hence the term "selective compliant." SCARA robots are widely used for assembly operations that require this rigidity in the X-Y plane, such as placing a round pin into a vertical hole without binding. An example of a SCARA-type Adept robot is shown in Fig. 4.11.

SCARA robots are less expensive than similar-size, fully articulated robots by virtue of the lower number of joints they possess. They are also faster and more compact than Cartesian gantry systems because the pedestal mount has a smaller footprint than the multiple-point mounting of a gantry.

FIGURE 4.11
SCARA robot.
(*Courtesy of Adept.*)

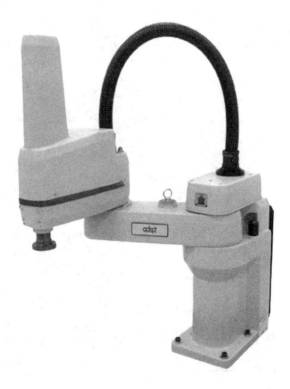

4.4.3 Cartesian Robots

A Cartesian robot, also called a linear or gantry robot, has three linear axes of control in the X, Y, and Z directions. Rather than having rotary joints, the X-axis is usually mounted at both ends with the Y-axis mounted to it. Some gantries suspend the Y-axis between two X-axes using a four post arrangement. This creates a box-shaped working envelope. The Z-axis is mounted to the Y-axis and may have an additional rotary axis mounted to the end. A gripper or other end effector is then attached to this for part handling. The Adept "Python" three-axis robot shown in Fig. 4.7 is an example of a Cartesian robot.

Gantry arrangements are the simplest control scheme for robots since coordinates are in the familiar X-Y-Z or Cartesian system and do not have to be converted or interpolated as with other systems. This allows for separate controllers or servo drives to be used for movements if coordinated moves are not required.

A popular application for Cartesian robots is the Computer Numerical Control (CNC) machine. These machines are used widely in industry for the automated machining of metal parts.

4.4.4 Parallel Robots

A parallel robot makes use of four or more linkages or kinematic chains from a central actuation point to an end effector. They are considered

Machine Systems

FIGURE 4.12 Parallel robot. (*Courtesy of Adept.*)

closed-loop systems since each of the links are constrained by the others. When compared with serial manipulators, such as robotic arms and SCARAs, the structural members are very light and therefore provide a much greater linear speed. The disadvantage of a parallel robot is that its work space is limited when compared to the space it occupies. Figure 4.12 shows an Adept "Quattro" parallel robot.

Parallel robots are usually suspended above the objects being manipulated. A common use for parallel robots is the insertion of components into printed circuit boards.

4.4.5 Robot Basics and Terminology

Robotic systems are made up of several components. The robot itself, with its motors, joints, and structures, makes up the moving part of the system. Motor and feedback cables are usually routed through the structural members of the robot for protection. Motors used in smaller robots are usually high-speed, low torque DC motors with high gear ratios. Larger robots use various different types of servomotors, depending on speed and payload requirements, but all use gearing or gearboxes of some kind.

For stability, the robot is mounted to a base, which is usually bolted to a solid foundation or steel frame. The base also usually has cable connections from the controller.

The robot controller contains drives for each of the axes along with the "brains" for running robot programs and coordinating axis movements. Communication ports for interfacing with programming computers and other controllers are also present. Safety interface connections for E-Stop and guard circuits are generally mounted here as terminal connections. Cables to the robot base connect to the controller as well as a port for the robot's programming pendant. This is also where the power connection for the robot is made.

The robot pendant is used to write or edit robot programs as well as to manually move the robot and "teach" positions. Pendants resemble an HMI with a touch screen and usually a membrane keypad. They also have an E-Stop button for integration into a machine safety circuit. A "dead-man" switch is built into the pendant, which must be held down during any manual robot movements.

The end effector is tooling placed on the working end of the robot. It is typically used to manipulate parts, but it may also be used to carry a welding tip or sprayer head. End effectors may carry pneumatic or hydraulic grippers, magnets, suction cups, or various other types of tooling. Some end effectors can be quite elaborate, having several actuators and sensors. Cameras and measurement devices are also often mounted on the end effector.

Terminology for robots and robotics can vary widely, depending on the manufacturer. Following are some of the more general terms that apply to a variety of platforms. Other terms for specific platforms can be found in manufacturers' documentation.

Specifications for a robot involve an analysis of the product and movements required. The *payload*, or carrying capacity, is the amount of weight that the robot must lift, including the weight of end effector tooling. This can be affected by the speed, acceleration, and force that are required. *Kinematics* is the arrangement of the rigid members and joints as described previously. Choosing between articulated, SCARA, Cartesian, or other configurations involves determining the envelope of the points that need to be accessed and at what angle. Two axes are required to access any point in a plane, while three are required to reach any point in a X-Y-Z space. In order to completely control orientation, an additional three rotational axes are required—pitch, yaw, and roll. Additional axes are sometimes added to reach around obstructions or into cavities.

Accuracy and *repeatability* are measurements of the precision at which the robot operates. Accuracy is a measurement of how closely a robot will move to a programmed or commanded position, while repeatability is a measure of how well the robot will return to the same position each time. *Compliance* is a measure of how much a robot will move when a force is exerted against it. When a load is being carried, positions will be slightly different than when there is no load at all. Joint positions may be detected accurately, but even solid members or joints will bend slightly under load. Acceleration can affect compliance even further and must be considered because of potential overshoot in position.

A *frame* or tool frame is a coordinate reference that is used to allow an entire set of points to be translated to a new location. As an example, if you have taught a large group of points around a station that the robot will access, then either the robot or the station needs to be moved. By using a frame recorded at X-Y-Z on the station, all that needs to be done is to teach the new frame, rather than reteaching all of the points.

A *singularity* is a condition at which robots can reach a point through more than one joint configuration, making the axes redundant. The American National Standard for Industrial Robots and Robot Systems—Safety Requirements (ANSI/RIA R15.06-1999) defines a singularity as "a condition caused by the collinear alignment of two or more robot axes resulting in unpredictable robot motion and velocities." It is most common in robot arms that utilize a "triple-roll wrist." This is a wrist about which the three axes of the wrist—controlling yaw, pitch, and roll—all pass through a common point.

The ANSI/RIA has mandated that robot manufacturers make users aware of singularities if they can occur while robots are being manually manipulated. Some industrial robot manufacturers have attempted to sidestep the situation by slightly altering the robot's path to prevent singularity conditions.

Often, SCARA robot arm positions are defined as being right- or left-handed to avoid singularities arising when the robot could access a point from either configuration. Articulated robots may have even more joint combinations that can theoretically reach a particular point through various arm configurations. Some singularity problems can be avoided by switching from Cartesian moves to individual joint movements within a program. Another method is to slow the robot's travel speed, thus reducing the speed required for the wrist to make the transition, avoiding the condition sometimes referred to as "wrist flip."

An area definition or vector can be used to define an area for safety or movement control. By defining a point along X, Y, and Z dimensions, a three-dimensional space can be used to control operations within an area. This is often used to set or reset an output when the robot enters or leaves a space. Terminology for this varies widely for different platforms.

An approach or depart move generally describes moving to or from a point at a certain angle. Rather than having to define a specific position for a move, the approach or depart move can make use of a defined position by telling the robot to move toward or away from a point using the current orientation of the end effector tooling.

4.4.6 Robot Coordinate Systems

Robot movements and positions can be defined in a number of different coordinate systems. "World" coordinate systems apply to any coordinate system using the robot's base as the origin. "Tool" coordinates use the end of the robot's arm where the tooling is mounted as the origin. "Workpiece" coordinates use a point on the work area tooling rather than a point on the robot itself as the origin.

The most familiar X-Y-Z, or Cartesian, coordinates are usually easiest for a human to visualize and so are often used for position definitions. Additional coordinates are sometimes added to the X (primary horizontal), Y (secondary horizontal), and Z (vertical). These are sometimes referred to as A (rotation around X), B (rotation around Y), and C (rotation around Z). SCARA and gantry-type robots require little interpolation or conversion to use this system. This coordinate system is sometimes referred to as a "space" coordinate system.

Joint coordinates describe the angular position of each of the robot's joints. Controllers use joint coordinates and perform mathematical calculations or interpolations to arrive at Cartesian points. These may be addressed in a program in variables such as J1-Jx or A1-Ax, depending on the software platform.

When operating a robot from a pendant, it may be advantageous to switch between different coordinate systems and work spaces, depending on the ease of visualization.

Further information on robot programming and software is addressed in Chap. 6.

CHAPTER 5
Process Systems and Automated Machinery

Process control is a method of manufacturing that uses formulas and recipes rather than assembling components to produce products. The primary way to determine if an application falls under the category of process manufacturing is whether the end product can be returned to its basic components. For example, an automobile can be disassembled and its components returned to stock, whereas a bottle of shampoo cannot be returned to its basic ingredients.

Process manufacturing is used in the chemical, food and beverage, pharmaceutical, biotechnology, and packaging industries. In the process industry, ingredients, formulas, and bulk are the elements of the end product rather than parts, assemblies, and components. Raw materials are also processed into intermediate forms that can be used in the manufacture of components. The materials that automated machines process all have their own special properties and techniques that are associated with their manufacture. There are many similarities in the processing of metals and plastics. For instance, both involve the mixing of different elements in a molten state and the machining or forming of shapes.

Automated machinery is created by combining the components and machine subsystems described previously in this book. Automated production lines use a combination of custom machines and original equipment manufacturer (OEM) equipment to assemble or produce an end product.

Some machines are fairly standardized, such as hydraulic presses, web-handling equipment, and injection molding machines. These machines are often made in several sizes and configurations, and only the tooling needs to be customized. They are often manufactured by OEMs who specialize in a specific kind of machinery. Other machinery is often customized to fit a specific application, such as assembly and gauging machinery. Custom machine builders often use a combination of OEM products and custom machines, moving components and products through a production or assembly line to produce the end result. The combining of the various machines and control systems into a single entity is a process known as *integration*.

5.1 Chemical Processing

Chemical processing involves combining or mixing ingredients and often changing their temperature or pressure. Some chemicals or compounds are produced in bulk form for use in further processing, including commodity chemicals in solid or liquid form, polymers or plastics, and petrochemicals. These products are generally packaged or contained in bulk for shipment to other facilities or processors.

Bulk chemicals may undergo further processing to produce specialty or fine chemicals, such as adhesives, sealants and coatings, industrial gases, electronic chemicals, catalysts, and cleaning compounds. They are also used in consumer products, such as soaps, detergents, lotions, and cosmetics. The life science industry also uses bulk chemicals and compounds in the production of pharmaceutical drugs and medicines, vitamins, and diagnostic products. Because of the higher research and development costs and government specifications and regulations, these products are usually produced in a controlled laboratory environment and are more costly.

Key process variables in the production of chemicals and compounds are residence time (lowercase Greek letter tau), volume (V), temperature (T), pressure (P), concentrations of chemicals ($C1$, $C2$, $C3$. ..., Cn), and heat transfer (h or U). Chemical processing is concentrated around the control and monitoring of these variables.

Chemical production and processing can be very hazardous because of the reactionary nature of chemicals. Pressures, temperatures, acidity, and quantities must all be monitored and controlled very precisely. This requires instrumentation and visualization using a wide range of products. HMIs are used in the field and in control rooms to display process diagrams and provide control and detailed alarms. P&ID diagrams are used to design and troubleshoot chemical processing systems and may also be accessed through computers or HMIs.

Valves for controlling the flow of liquids and gases in a process are often analog; they not only completely open or close, but also move to intermediate positions. These are known as *proportional valves*. They also usually have a feedback sensor to verify position, although sometimes a downstream flow sensor or pressure feedback is used. Process variables are monitored using standardized limits such as H and L, indicating values within which operation is considered normal, and HH and LL, indicating limits that require alarms and may initiate shutdown.

Because of the caustic and explosive nature of many chemicals, safety is an overriding concern in the chemical industry. Controls are often redundant and mechanical elements are designed with high safety margins. IS and explosion-proof products are used extensively within chemical process facilities. Both PLCs and DCSs are used for control along with stand-alone temperature and PID controllers. The physical layout of chemical processing facility generally includes elaborate piping with multiple holding and containment vessels.

Vessels designed to contain chemicals as they undergo mixing or property changes are known as *reactors*. These are designed to maximize their value for the given reaction, providing the highest yield while minimizing cost of materials or energy. Basic vessel types include tanks in various shapes or pipes (tubular reactors). They may operate in a steady state, where materials flow into and out of the system continuously, or a transient state, where a process variable such as heat or pressure is changed over time.

Three main basic models for estimation of process variables are used in chemical processing; the batch reactor model (batch processing), the continuous stirred tank reactor model (CSTR), and the plug flow reactor model (PFR). Catalysts may require different models or techniques than these three basic models, such as a catalytic reactor model.

5.2 Food and Beverage Processing

Food processing uses meat, grain, or vegetable components to produce packaged food products for commercial use. Like chemical processing, food processing involves the control of temperature and often the mixing of ingredients. Administration of the food-processing industry regulations in the United States is monitored by the Food and Drug Administration (FDA) and the United States Department of Agriculture (USDA). Of primary concern is sanitation and the elimination of contamination.

Special techniques specific to the food industry include clean-in-place, flash freezing, spray drying, and various filtration methods. Water treatment is also an important part of the process because of the necessity of cleaning equipment with caustic substances and federal regulations involving wastewater discharge. Product or material handling and packaging are also integral to the food-processing industry.

Food-processing machinery may be made by OEMs who specialize in specific aspects of production or by custom machine builders who are knowledgeable of the special requirements of food processing. Food-processing machines may be used for the preparation of ingredients (chopping or cutting, crushing, peeling, or molding into shapes); the application or removal of heat (cooking, baking, or freezing); mixing of ingredients or seasoning; or product filling. Most food service equipment is built using stainless steel and techniques allowing for pressurized wash-down of equipment. Great care is taken to ensure equipment does not have crevices where contaminants can lodge.

Controls in the food- and beverage-processing industries are similar to that of chemical processing, although mostly PLC controlled. Instrumentation is used to measure temperatures and flow rates, sometimes using individual controllers with data read back into a SCADA or monitoring system.

Food is produced by several different methods. *One-off* or individualized production does not lend itself to automated methods. An example of this would be wedding cakes or cake decoration. *Batch* production is commonly used in bakeries, where a certain number of products of the same size and ingredients are made on a periodic basis. Equipment is set up with an end number in mind, and ingredients are ordered based on an estimate of the demand. Many OEM food production machines are constructed with recipe, batching, and counting features built into the software. Weighing and liquid measurement are also an important part of the batching process. *Mass production* is a continuous method of producing products, such as canned or packaged foods, and individual items, such as candy. Product passes from one stage to another along a production line in this method.

Beverage processing involves the formulation of the product as well as the bottling or packaging. Basic ingredients may be mixed in a batch form to control proportions accurately or in a continuous manner. Some beverages, such as beer or whiskey, require long periods where ingredients age or ferment in a tank at a specific temperature. Others can be continuously processed directly into a bottling or packaging area. Bottling is a high-speed process with many OEMs involved in making standard or semicustomized machinery. Packaging equipment, from sterilized product containers to bulk shipping, is also a major part of both the food- and beverage-processing industries.

Like the chemical-processing industry, food and beverage processing involves a great deal of process visualization. HMIs and integrated control systems allow the viewing of the entire process from raw ingredients through packaging. Many control component vendors have packaged templates specific to the food and beverage industry with three-dimensional images, recipe management, and historical data collection. There are also quite a few integrators and custom machine builders who specialize in food- and beverage-processing equipment.

5.3 Packaging

The packaging industry encompasses the containment, labeling, orientation, and material handling of products for distribution, storage, and sale. Products are packaged in different ways, depending on the stage of production and type of product. Many products are produced in bulk form but must be packaged as individual units for shipment or sale.

Packaging can be described in three broad categories. Primary packaging is the first layer of material that surrounds or holds the product. This layer is often consumer packaging that is labeled for marketing and comes in direct contact with the product. Secondary packaging is often used to group primary packaging together, usually also labeled for consumer use. Tertiary packaging is used for bulk handling, storage, and shipment.

Packaging materials are generally made of some form of plastic or cardboard, although in the food and pharmaceutical industries, glass and metals are also widely used. Packaging machinery generally takes these materials in a rolled or collapsed form and uses it to envelop the product.

Common methods of primary packaging include shrink-wrapping, cartoning, blister packing, and vacuum packing. These methods also lend themselves to easy labeling for consumers by application of adhesive labels, direct printing, or printed cartons and bags. Secondary methods also include cartoning, bagging, and shrink-wrapping. Tertiary packaging uses corrugated cardboard, palletizing, bag-in-box, and other techniques. Packaging for transport is typically less concerned with marketing and the appearance of a packed unit than with protection and ease of handling.

Packaging machinery can be purchased as off-the-shelf equipment from many OEMs. Labelers, check-weigh systems, baggers, case erectors, sleeve-wrappers, and shrink-wrappers are made in standard sizes by many manufacturers and can be ordered with short lead times. They are usually adjustable for various standardized sizes of packaging and labeling materials. Machinery can also be customized by many manufacturers from a standard design. Some packaging machinery must be custom-built because of special requirements related to material handling, package size, or speed requirements. End users often customize or manufacture their own packaging machinery in-house.

Considerations when choosing packaging machinery includes type of packaging and its final appearance, floor space requirements, throughput, reliability and maintainability, labor requirements to operate machinery, and flexibility of the equipment related to product sizes and changeover. Packaging machinery is often built into a line involving material handling and an integrated control system. Accumulating, orienting, and collating are commonly part of the material-handling process as well as inspecting and weighing of product. Machine vision and metal detecting are common in the packaging industry. Figure 5.1 shows a typical packaging line layout.

Liquid packaging includes filling, capping, closing, seaming, and sealing. Because the packaging of liquids may involve foods or beverages, cleaning and sterilizing may also be part of this process. Cooling and drying are also common in the material handling and packaging of products.

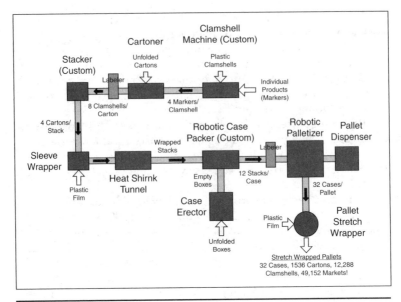

FIGURE 5.1 Packaging line layout.

The use of plastic wrap involves shrinking film using ovens and heated air and the sealing of packages using bar sealing equipment. Temperature and time are important variables in this process and are often controlled by discrete devices, such as temperature controllers, timers, and variable speed conveyors. Packaging machinery uses a wide range of technologies ranging from simple control and mechanical components to high-speed servo and robotic systems.

Palletizers are used to arrange boxes and cartons in layers on plastic or wooden pallets. They are often robotic and can be programmed for different patterns. Pallets are then usually stretch-wrapped using a rotary wrapper or by hand and transported by forklift to the shipping area.

Labeling may be applied to packages by means of adhesive labels from a dispenser or directly printed onto cartons, bags, or boxes. Labelers are often controlled by means of a part detection sensor acting as a trigger and may also use an encoder or other speed detection device to control spacing. Bar coding is also an important part of the labeling industry. Bar codes may be preapplied to labels or printed directly on the product.

5.4 Web Handling and Converting

A *web* is a continuous piece of material. Paper, fabric, and some extruded materials are examples of a web. Webs are usually moved through processes using rollers, which may also serve to apply tension, heating, or cooling to the material. Webs may be processed directly from the manufacturing of the material or from rolls or coils. Common materials processed in web form include Nonwovens, films, textiles, foams, and paper.

Because webs are moved continuously, they can be processed at higher speeds than individual sheets of material. Common uses of web processing include bonding layers of material together; cutting material into sheets or slitting it into smaller rolls; cutting or punching pieces out of it; or passing the material through other heating, cooling, or chemical processes.

Web-handling machinery is usually composed of individual drives arranged in a linear fashion with sections performing different operations on the web sequentially. Web guiding and tensioning are important facets of controlling webs, and a variety of products are manufactured to help in these areas. *Web guides* and steering mechanisms sense the position of the edges of the web and adjust rollers in real time to correct for misalignment. These are often placed immediately before critical sections of a line, such as printing. Tension may be monitored by load cells and strain gauges, or torque feedback may be sensed by roller drives. Gaps between rollers must often be controlled and adjusted carefully for tensioning or product thickness. *Pull units* are drive and roller sections that have methods for these adjustments built in. They may consist of full-width or "nip" rollers at the web edges. Figure 5.2 shows a web being processed.

Accumulators are used to isolate sections of web from the tension effects of other sections. They also store material while rolls are being changed. Accumulators are sometimes built with multiple festoons or may take the form of a simple take-up mechanism. *Unwinders* and

FIGURE 5.2   Web line.

winders pay out rolls of material and wind them up at the ends of the process. *Splicers* are placed between the unwinding station and an accumulator to allow web to feed out continuously. These stations may be fully automated or operator assisted.

Slitters are used to cut webs lengthwise on a continuous basis. They may have stationary blades or rolling knives and are usually adjustable for different widths.

Some operations, such as heating, cooling, and slitting, are continuous processes, while other operations, such as ultrasonic welding, punching, or cutting across the web, may be done by starting and stopping or slowing the web between accumulators.

Converting is the process of turning raw materials, often in rolled or sheet form, into new products. Common materials used in the converting process are woven and nonwovens, paper, adhesives, rubber, foam, and plastics.

Continuous rolls of material are threaded into processing machinery, which then performs intermediate or final operations to produce product. An example of converting is taking a web of two layers of plastic, fusing the edges together and sealing the ends, and cutting it to produce plastic bags. Common operations in the converting process include coating, application of adhesives, ultrasonic welding, or other types of material bonding, sealing, and patterning. Converting methods are also often used in the assembly process.

5.5 Metal, Plastic, Ceramic, and Glass Processing

Processing of metals, plastics, ceramics, and glass are similar in several ways. They all involve the combination of raw materials and often the application of heating, cooling, chemicals and pressures. They are processed in their solid and liquid, or molten states. They often pass through several intermediate forms and shapes before being made into a final product.

5.5.1 Metals

Raw metals are extracted from the earth in the form of ores that must be processed to extract the pure metals. The initial processing may use chemical or electrolytic reduction, pyrometallurgy (high temperature), or hydrometallurgy (aqueous or water-based chemistry). When an ore is an ionic compound of a metal with impurities, it must be *smelted* to extract the pure metal. Ores such as iron, aluminum, and copper are typically mixed with other compounds or chemical elements, which must be separated out by breaking the bonds either electrically or chemically. Many common metals, such as iron, are smelted by combining the ore with carbon as a reducing agent at high heat.

To separate aluminum from the ore bauxite—a common practice in extractive metallurgy—carbon and electricity must be introduced.

The aluminum is extracted using an electrochemical process in a carbon-lined vat or "cell" using molten cryolite or synthetic sodium aluminum fluoride. Aluminum smelters consume a great deal of electricity because of the very high melting point of the metal.

Copper ore contains a very low percentage of copper metal and goes through several stages to purify the metal. The ore is first ground and separated from other minerals. It then undergoes hydrometallurgical or froth flotation procedures to further refine the metal before being smelted. The smelter produces about a 70 percent copper sulfide, which is then further refined and purified using electrolysis.

Purified metals or alloys are usually made into a solid form, such as ingots, sheets, or coils, and shipped to other facilities for further processing.

Alloys

An alloy is a mixture of materials in which the major component is a metal. Alloys of iron are the most common, including tool steel, cast iron, and stainless steel. Alloying iron with different amounts of carbon produces low-, mid-, and high-carbon steels. More carbon reduces the ductility but increases the hardness and strength of steel.

Cast iron is an alloy containing iron, carbon, and silicon. It is melted and poured into molds to form different shapes, which are then often processed further. Common uses for cast iron include motor housings, pipes, and machine parts. Cast iron can be brittle, so other elements are often alloyed with it to increase its malleability, hardness, or tensile strength. Cast iron has a relatively low melting point and is easy to machine.

Stainless steel is made by alloying carbon steel with chromium. Other elements, such as nickel and molybdenum, are sometimes added to make it less brittle or increase hardness. Stainless steel is used primarily for its resistance to corrosion. It is used in food processing and medical devices as it can be easily sterilized and does not need painting or other surface coatings. Stainless steel is commonly formed into coils, sheets, plates, bars, wire, or tubing before being further processed.

Copper, aluminum, titanium, and magnesium are other common alloys produced for commercial use. Copper and its alloys are often used for electrical wiring, while alloys of aluminum, titanium, and magnesium are valued for their high strength to weight ratios. Alloys designed for special or very demanding applications may contain more than 10 elements.

Smelters are major users of automation. Because of the high temperatures and toxic fumes, it is difficult for operators to work in close proximity to the molten materials. Material handling, power distribution, and process control are important elements of metal alloying and refining. Visualization of the process using HMIs and SCADA (System Control And Data Acquisition) are common.

Metal Processing

Once a metal has been alloyed into its final constitution, it must be brought into a useful form. It is common to melt the raw alloy into ingots for further treatment or use a continuous process to form it into sheets or plates. This is often done at the smelter prior to shipping using extrusion or rolling mills.

Casting is a forming process that requires melting a metal and molding it into a shape. There are a variety of methods to accomplish this. Molten metal may be poured directly into a form using investment or "lost wax" casting. Metals may also be forced into a die at high pressure (die casting), poured into a mold made of sand (sand and shell casting), or formed by rotating the molten materials within a mold (centrifugal, spin, and rotocasting).

Extrusion uses metals in a liquid, plastic, or solid state to shape it by forcing it through a die. The metal is pushed or drawn at high pressure through an opening of a desired shape. It is then often stretched to straighten it. The process may be continuous or produce individually shaped pieces using a "blank" or billet. Extrusion presses may be hydraulically or mechanically driven.

Wire is usually made by drawing annealed metal through a hole in a die. Dies used for producing wire are often called draw plates and may have more than one hole for pulling multiple wires. Annealing, a heat treating process, makes metals or wires more flexible or ductile. Wire is often drawn more than once, making it thinner each time. It can be drawn up to three times before it needs to be annealed again.

Thicker cables or stranded wire is woven or twisted from individual wires by specialized equipment, often in machinery located directly after the draw plate. Wire is also woven into wire netting or wire cloth and mesh and is often further treated with heat or chemicals to change its properties after being drawn. Common metals used for wire are copper, aluminum, silver, platinum, iron, and gold, which have the ductility that make them suitable for wire.

Rolling is a technique that passes metal stock through a pair of rollers. This may be done at high temperatures, above the recrystallization temperature of the metal, or at lower temperatures; processes known as hot rolling and cold rolling, respectively. Metals may be rolled into rectangular cross sections as sheets or plates, rolled into a very thin sheet called a foil, or passed through consecutive rolls to shape the cross section, a process called roll forming. Roll forming is usually performed on coiled steel rolls. Figure 5.3 shows a roll former with progressive forming dies.

If metals are rolled thin enough, they may then be rerolled for shipment. Larger rolls are often slit into smaller rolls by other facilities.

FIGURE 5.3 Roll former. (*Courtesy of Mills Products.*)

The smaller rolls may then be further processed using cutting, punching, or pressing techniques.

Forging uses pressure to form metals into desired shapes. Like rolling, forging may take place at temperatures above or below the recrystallization temperature of the metal, known as hot or cold forging. Cast or formed pieces may be further processed after the shape is cooled. They are usually finished using presses or machine tools.

Drop forging uses a heavy weight or "hammer" to strike and deform the workpiece. Drop forging uses open or closed dies to shape the material. Open die drop forging is often called smith forging, while closed die forging is known as impression forging. In impression forging, the workpiece is often moved through a series of cavities in the die to achieve its final form. The hammer may be dropped several times, depending on the complexity of the part being forged.

Press forging applies a continuous pressure to the metal as it is formed. Like drop forging, dies can be partially open, allowing material to flow outside the die, or completely closed. Presses are typically hydraulic when used for forging. Press forging allows for more complete control of the forging process since the rate of compression can be timed.

Another form of press forging is known as upset forging. This is a method of increasing the diameter of a workpiece by reducing its length. Dies in upset presses often have multiple cavities and are used to compress rods or wires (thin rods) into the raw stock for bolts, screws, and other fasteners. The raw stock is then threaded in other machinery. Engine valves and couplings are other products often produced by upset forging.

Presses are also used on sheet metals to form shapes by drawing or stretching it. If the metal is stretched longer than its diameter, it is known as deep drawing. Pressure and lubrication are important variables used in the drawing process, along with the rate of die movement. Another common use of presses is to punch shapes into sheet metal or to "coin" parts, impressing a pattern onto the surface.

Presses often have self-contained control systems. Hydraulic presses have pumps to move hydraulic fluid through cylinders and often also have a coolant pump. Motor starters or variable frequency drives are usually part of the control panel. Proportional valves and positional feedback are also typical elements of a press. Typically, presses are controlled by a PLC also handling safety and other actuators and sensors. Figure 5.4 shows a hydraulic press with the die removed.

FIGURE 5.4 Hydraulic press. (Courtesy of Mills Products.)

Hydroforming applies water at high pressure to the inside of a closed metal shape such as a pipe. This allows the forming of contoured shapes such as curved handles. Pressurized liquid is introduced into the shape during the final closing of a hydraulic press die.

Powder metallurgy (P/M) is the compression of metal powder into a form of the desired shape. A number of different metals and their alloys are used in the P/M process, including iron, steel, stainless steel, copper, tin, and lead. Powders are produced using atomization (transforming a stream of molten metal into a spray of droplets that solidify into powder), chemical methods, and electrolytic processes. Powders can then be blended and ground further in a ball mill.

Forming of components is done using a variety of pressing techniques. Mechanical presses such as crank and toggle presses are often used if a high production rate is required; however, forces are limited. Hydraulic presses are used for higher loads, but production rates are lower. The formed component is then heated (but not melted) to allow a metallurgical bond to form between the particles. The forming or compacting process is usually done at room temperature in a die. The result of this operation is a briquette or green compact with a solid form. This shape is quite fragile and must be handled carefully.

The second step in the formation of a P/M component is the heating of the component, typically in a hydrocarbon gas atmosphere. Formed briquettes are usually heated to between 60 and 80 percent of the melting point of the constituent with the lowest melting point. This process is called *sintering*. The result is a formed metal solid that is smaller than the original component. Secondary operations such as restriking or reforging, sizing, heat treatment, impregnation, machining, grinding, and finishing may also be part of the process. Small parts are often made using P/M techniques because the density of the metal is more uniformly controlled.

Other metal-processing methods include various cutting and machining techniques. Shearing, sawing, and burning are common cutting methods. Erosion methods such as water jet or electric discharge are also used to cut metals. Wire electrical discharge machining (EDM) uses electrical current to erode material and allows detailed cutting beyond the capability of chip-producing processes.

Machining methods are chip-producing processes used to shape metals. One of the most common machining techniques is drilling, which uses the point of the tool to make holes in material. Milling uses either the side of a tool or its face to remove material. Grinding and polishing are other machining methods. Turning uses the rotation of a workpiece and a fixed or slowly moving tool to remove material. A common machine using this technique is a lathe. Where heat is generated by tool friction, cutting fluids or coolant are often used to remove heat from the workpiece. Coolant is usually recaptured, filtered, and cooled before being reused on the product.

Machine tools are often automated; servo systems precisely control speeds and positions. Computer numerical control (CNC) machines allow CAD and CAM programs to be interpreted and move tooling automatically to produce a part. Tools are often automatically changed using robotic "tool changers." Robotic operations may also be used to move parts from station to station within a workcell.

CNC machines are usually programmed using G-code; however, other languages such as Step-NC and control languages similar to BASIC are commonly used also.

After forming metals into the desired shape, the metal is often treated using chemicals, heat, and electricity to change its properties. Heat treatments are generally used to change the ductility, malleability, or hardness of metals, while chemicals and coatings are used to change the finish or corrosion resistance of metals. *Galvanization* applies a protective coating of zinc to steel or iron to prevent rusting. *Anodization* increases the thickness of the natural oxidization layer on metal parts using an electrical process, which improves corrosion resistance and provides a good surface for adhesives or paints.

Like the smelting industry, metal processing uses automation extensively. Thicknesses and lengths must be accurately measured and roll speeds controlled and monitored. As with other industries, there are many OEMs that manufacture equipment specific to metal processing. Machines are often built around standard equipment to provide material handling and safety features. Production lines are built using individual machines, often taking product from a raw sheet, ingot, or blank to its finished form.

5.5.2 Plastics

Plastics are made from organic materials, usually synthetic or semisynthetic solids. Crude oil is processed using a method called catalytic cracking to break it down into substances such as gasoline, oils, ethylene, propylene, and the butylenes. Natural gas is processed using thermal cracking to produce many of the same components. The petrochemical monomers or raw materials that are derived from these substances include ethylene glycol, isobutene, isopropyl benzene, toluene, chloroprene, styrene, and many more. These raw materials are then processed further to produce rubbers, adhesives, lubricants, asphalts, and plastics.

There are two categories of plastics: thermoplastics and thermosetting polymers. All plastics are formable when they are heated, thus the prefix *thermo*. *Thermoplastics* do not change their chemical properties when they are heated and can be formed or molded multiple times. Polyethylene, polypropylene, polystyrene, polyvinyl chloride (PVC), and polytetrafluoroethylene (PTFE) are all examples of thermoplastics.

Thermosets or thermosetting polymers are formed into a shape one time. After they are formed, they remain solid. Thermosets are usually liquid or malleable before they are cured by heat or chemical processing. They are usually stronger than thermoplastics but are also more brittle. Epoxy resins, Bakelite, vulcanized rubber, Duroplast, and polyamides are examples of thermoset materials, as are many adhesives.

Additives are used to change the hardness, color, flammability, biodegradability, or other properties of the plastic. These may be added during the preprocessing stages of creating the plastic or added later while forming the part.

Extrusion

Plastics extrusion is a process that forms plastic material into a continuous profile. Extrusion is used to produce pipe and tubing, adhesive tape, wire insulation, and various plastic framing profiles.

Raw thermoplastic material in the form of beads or pellets is fed from a hopper into the rear part of the extruder barrel. A rotating screw forces the beads, also called resin, forward into the barrel. Additives are sometimes mixed with the resin to color the plastic or make it UV resistant. The barrel is heated to melt the plastic, usually to between 200°C (392°F) and 275°C (527°F), depending on the polymer. Heating is often done in stages, using different controllers for each stage, allowing the beads to melt gradually. Pressure and friction within the barrel contribute to the heat of the melt. Sometimes air or water is used to cool the polymer if the material becomes too hot. Figure 5.5 shows a thermoset material being extruded from a die.

After the molten plastic leaves the barrel, it is passed through a screen pack and breaker plate to remove contaminants. It then enters the die, where it is shaped to the desired profile. After exiting the die, the material is pulled through a cooling section, typically a water bath. Plastic sheeting is sometimes fed through cooling rolls. After cooling, secondary processes may be performed, like applying adhesive to tape. The finished product is then cut into sections, rolled, or spooled.

Injection Molding

Plastic injection molding is a process that can be used for both thermoplastics and thermosets. The initial stages of injection molding are similar to that of plastic extrusion, with beads or resin being fed into a heated barrel from a hopper. It is then injected into a mold cavity, where it cools and hardens in the shape of the mold. Molds are usually made of steel or aluminum and are precision machined to form the features of the part.

Thermosets typically use two different materials that are injected into the barrel. The materials begin chemical reactions that eventually harden the material irreversibly. This can cause problems if the

Chapter Five

FIGURE 5.5 Extrusion.

material were to harden inside the barrel, so minimizing the time the material is in the barrel and screw is important.

Molds often have channels between cavities to allow flow of material, called sprue, which must be removed from the mold and the product. The mold also needs to be separated to remove the part. This separation, along with ejector pins that help push the part out of the mold, creates lines and marks on the product. Material that needs to be removed from the product is called flash and must often be removed manually. Figure 5.6 shows a commercially available injection molding machine.

Injection molding is a common process for manufacturing parts of every size. Injection molding machines are usually manufactured by OEMs who specialize in plastics. They are usually PLC controlled and are configurable through an HMI. Recipes and control of temperatures, times, and speeds are standard features. Automatic unloading of parts is often a feature of injection molding machines.

Thermoforming

Plastic sheets or films can be formed into a mold by heating the material and pulling it into a mold with vacuum. This process is

Process Systems and Automated Machinery

FIGURE 5.6 Injection molding machine. (*Courtesy of Hope Industries.*)

called thermoforming. Material must then be trimmed from the edges after cooling. Often, secondary processes such as drilling or punching are performed while the part is still on the machine. This and the trimming operation can be incorporated into the control system if required.

Automated thermoforming machines are much simpler to manufacture than extrusion or injection molding machines. They are often made by custom machine builders and machine shops rather than OEMs since they consist of essentially a frame to hold the mold, heaters, and plumbing for vacuum. If water or air cooling is necessary, this is usually easy to implement also.

Thermoforming is used on products such as cups, lids, and trays for the food industry; blisters and clamshells for the packaging industry; and specialty components for the medical field. All of these are examples of thin gauge thermoforming. Large production machines making thousands of parts an hour are often used for thin gauge products. Parts can be formed continuously at high speed using a sheet or film on a roll, similar to web processing. Indexing chains are often used to transport the material across molds and through ovens and cooling zones. These chains may have pins or spikes to hold the material between the molding areas, helping in the transport process. The remaining material is rolled up at the end of the line after formed parts have been removed. This material can then be recycled by grinding it up in a granulator.

Thick gauge products are usually made at a much slower rate because of the longer heating and cooling times. Unlike thin gauge forming, thick gauge processing usually involves loading and indexing individual sheets through multiple stations. Sheets are placed or clamped into a frame at a load station. Stations may be indexed linearly or on a rotating carriage. The first station heats the sheet, usually in a pressure box with mating molds closed over the sheet. Vacuum and pressurized air are used to pull or push the sheet into the contours of the mold. Some molds will have movable sections that help push the material into parts of the mold with an actuator; this is called a "plug." Plug assists are used for taller or deep-drawn parts in addition to vacuum to help distribute the material evenly.

After a defined forming time, the mold is opened. At the same time, the vacuum is removed and a burst of air is applied to the mold—an action known as air eject. A stripper plate may also be actuated to assist in the removal of the formed sheet. The formed sheet is then indexed into a station that cools the sheet and cuts the parts out of the sheet with a die. Automotive door and dashboard panels, plastic pallets, and truck bed liners are examples of thick gauge thermoforming.

Blow Molding

Hollow plastic items, such as bottles, are formed in a process called blow molding or blow forming. There are several techniques used in this process, including extrusion blow molding (EBM), injection blow molding (IBM), and stretch blow molding (SBM). Deformable or moving dies are also sometimes used in the process.

The blow molding process begins with a plastic form called a preform, or parison—a tube of plastic with one end open for air injection. The parison is clamped into a mold, the parison is heated, and compressed air is blown into the opening inflating the form into the shape of the mold. After the plastic has cooled and hardened, the mold is opened and the part ejected. Thicker parts often have excess flash still attached to the part that must be trimmed. For cylindrical parts, this is usually done using a spin trimmer, which rotates the part while trimming material with a titanium blade. Figure 5.7 illustrates the blow molding process.

FIGURE 5.7
Blow molding.

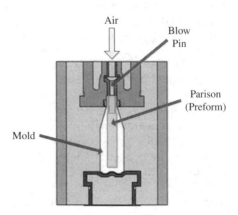

Stretch molding first uses a preform that often has premolded features, such as a threaded neck for a cap connected to the parison. These preforms are then reheated in "Reheat Stretch" blow molding machines. The stretching of the plastic induces strain hardening of the plastic, which is useful for bottles that contain carbonated beverages.

EBM is used on continuous profile such as hoses and pipes. This is done in much the same way as standard plastic extrusion, although air is blown around a mandrel through the center of the tube.

Common plastics used in the blow molding process include PET, PC, PVC, HDPE, PE+LDPE, and LLDPE. Other brand-name products such as BAREX and K Resin are used also. Most blow molded materials are thermoplastics.

Plastic Sheet and Materials

Plastic sheet and film is produced by a rolling process called *calendering*. Material is formed by passing it through a series of heated rollers. Thickness is determined by setting the spacing between the calendering rolls. For final packaging and shipment, film can be rolled or spooled after cooling, while thicker plastic sheets are cut and a protective removable film is often applied before it is stacked and ready for shipment.

Sheet, films and machinable plastics are used in the packaging and machine-building industries, so it is important to consider some of the properties of plastics.

The following information is compiled from the US Plastic Corp. knowledge base:

Acrylonitrile-butadiene-styrene (ABS): This thermoplastic material has good impact strength, formability, stiffness, and toughness. Good chemical and stress-cracking resistance. A good general-purpose, low-cost material. Easily thermoformed, strength is affected by temperature. Black is UV resistant, while white and natural colors are not. Applications include aircraft interior trim, tote bins and trays, cassette holders, automotive parts, and luggage. The maximum working temperature is 185°F, forming temperature 325 to 350°F.

ACETAL (Delrin): Excellent load-bearing qualities in tension and compression. Does not absorb a large amount of moisture. High-yield strength at elevated temperatures. Machinable, easily fabricated. Low-friction, high-wear resistance. Attacked by strong acids and oxidizing agents, resistant to a wide range of solvents. Not UV stabilized. Excellent material for bearings, gears, cams, and small parts. Meets FDA standards, USDA approved. Service temperature range 20°F to 185°F, intermittent 200°F.

Acrylic: Completely transparent, flexible, resistant to breakage. Lightweight (half the weight of glass), virtually unaffected by exposure to nature, salt spray, corrosive atmospheres. Easy to fabricate, can be sawed with fine-tooth blades, drilled with plastic drills, sanded, and polished. Can be cemented with acrylic cement. Meets FDA standards, UV stabilized, UL 95 flammability rating. Used for inspection windows, sight gauges, windshields, meter faces, protective covers, safety shields, tanks, trays, and displays. Service temperature range –40°F to 180°F, forming temperature 350°F.

CPVC (Chlorinated Polyvinyl Chloride): Boosts working temperature of other rigid vinyl thermoplastics by 60°F without affecting corrosion resistance. Safely handles many corrosive liquids. Virtually immune to solvents and inorganic reagents, aliphatic hydrocarbons, and alcohols. Corrosion resistant, lightweight, high tensile strength, noncombustible, low flow resistance. Not UV stabilized. Service temperature 33°F to 212°F, forming temperature 310°F to 325°F.

Nylon: High wear and abrasion resistance, low coefficient of friction, high strength to weight ratio. Corrosion resistant to alkalis and organic chemicals. Nonabrasive to other material, has noise-dampening characteristics and is a good electrical insulator. Not UV stabilized, USDA and FDA compliant. Used for bearings, bushings, washers, seals, guides, rollers, wear plates, fasteners, insulators, forming dies, sleeves, liners, cooling fans. Temperature range –40°F to 225°F.

Polycarbonate (Lexan): High-impact material, virtually unbreakable. UV stabilized, can be sawed with fine-tooth blades, drilled with plastic drills, sanded, and polished. Applications include greenhouses, window glazing, safety guards, chair mats, equipment enclosures, signs, and doors. Continuous service range –40°F to 240°F.

Polyethylene LDPE (Low Density): Semirigid material with good impact and abrasion resistance. Excellent corrosion resistance to a wide range of items. Susceptible to stress cracking when exposed to ultraviolet and some chemicals; wetting agents and detergents accelerate this. Can be heat formed, shaped, and welded to fabricate ducts, hoods, and more. Cannot be cemented but easily welded using a plastic welder. Can be cut with a wood saw and regular metal bits. Not UV stabilized but meets FDA standards. Working temperature of 0°F to 140°F, forming temperature 245°F.

Polyethylene HDPE (High Density): Rigidity and tensile strength of HDPE resins are much higher than LDPE and MDPE; impact strength is slightly lower. Rigid with good abrasion resistance. Other features and uses similar to LDPE as far as uses and workability. Working temperature –60°F to 180°F, forming temperature 295°F.

Polypropylene: Good balance of thermal, chemical, and electrical properties with moderate strength. Hard, high-gloss surface is suitable for environments that have a concern for bacteria buildup that can interfere with flow. Other features and uses similar to LDPE and HDPE as far as uses and workability. Not UV stabilized but USDA approved and meets FDA standards. Working temperature of 32°F to 210°F, forming temperature 310°F to 325°F.

Polyurethane: High load-bearing capacity and excellent tear resistance. Abrasion, oil, and solvent resistant and a high resistance to sunlight and weather conditions. Provides superior sound-dampening properties relative to rubber and plastics. Good electrical insulator. Applications include gaskets, seals, gears, wheels, bearings, bumpers, drive belts, valve seats, noise damper, chute hopper liners, mallet heads, and solvent lines. Working temperature range −90°F to 180°F.

Polyvinyl Chloride (PVC): Excellent corrosion and weather resistance. Good electrical and thermal insulator. Self-extinguishing per UL Test 94. Not UV stabilized, not FDA approved. Applications include corrosion resistant tanks, ducts, fume hoods and pipe, fabricated parts, tank linings, and spacers. Working temperature of 33°F to 160°F, forming temperature 245°F.

PVC Expanded Sheet (Sintra): Moderately expanded high-density, foamed material approximately half the weight of PVC. Easily cut, sawed, drilled, and fabricated. Can be painted and silk-screened. Durable and hard wearing, resists most chemicals and water. Fire retardant, sound dampening. USDA approved, recommended for indoor signage and displays.

Styrene: High-impact resistance, dimensionally stable. Low water absorption, heat and electrically sealable. Nontoxic and odorless. Can be painted and has good forming properties using vacuum pressure. Can be drilled, threaded, sawed, sheared, punched, and machined. Used for models, prototypes, signs, displays, enclosures, and more. Not UV stabilized but meets FDA standards. Maximum heat resistance 180°F, forming temperature 325°F to 350°F.

Teflon: Nearly impervious to chemicals; only molten alkali metals and gaseous fluorine at high temperatures and pressures attack it. Lowest frictions coefficient of any solid, no slip-stick characteristics (static and dynamic coefficients are equal). Nothing sticks to unheated surfaces. Virgin and mechanical grades UV stabilized; virgin material meets FDA standards and is USDA approved. Working temperature range is −20°F to 500°F.

Ultra High Molecular Weight Polyethylene (UHMW): Exceptionally high abrasion and impact resistance. Will outwear metals, nylons, urethanes, and fluoroplastics. Corrosion resistance is similar to other polyethylenes. Self-lubricating, nonadherent surface. Applications include guide rails, wear plates, rollers, conveyor augers, bin and hopper liners, chutes, bearings, bushings, and gears. FDA and USDA concurrence for contact with food and drugs. Working temperature range is −60°F to 200°F.

Composites and Reinforced Materials

Composite materials are plastic resins that have been reinforced with organic or inorganic materials. They differ from reinforced materials in that the fiber structure of the reinforcing material is continuous.

Plastic resins may be reinforced with cloth, paper, glass fibers, and graphite fibers. Usually these fibers are in short pieces that are randomly oriented since they are simply mixed with the resin. They do not have the same strength as true composite materials, which run unbroken through the resin. Reinforced plastics are extremely strong, durable, and lightweight. Reinforced plastics are manufactured in several different ways, including calendering of cloth with a plastic coating. Plastic coatings can also be brushed or sprayed onto materials.

Laminating is the process of layering different materials together. Often done at high pressure, sheets of material are placed between heated steel platens and compressed hydraulically. This bonds the layers of a laminate into a rigid sheet. These laminates can also be formed around corners and bonded to other materials.

Another common process is fiberglassing. Alternating layers of glass fiber fabric and plastic resin are coated over a form or mold. While the resin is in the liquid state, short pieces of glass fiber can be mixed in to provide added structure. After hardening, fiberglass can be finished using mechanical abrasives, such as sanding and buffing. Common uses for fiberglass are in the manufacture of boat hulls and swimming pools.

Composites are very lightweight and durable. A popular composite material is epoxy resin reinforced with graphite fibers. The fibers form the main structural component of the composite, usually the reinforcing fiber, makes up about half of the total material weight. Both thermoplastics and thermosetting materials are used in composites. Thermosets are often used because of the high heat that composites must sometimes withstand.

Composites can be manufactured in several ways. One method is a process called reverse extrusion, or "pulltrusion." In this method, the fiber portion is pulled or drawn through liquid resin and then through a heated die. Structural members and tubing can be made in this manner. Another method is winding filaments back and forth around a form and coating the fibers with epoxy resin. This method is used to form hollow products like tanks and pressure vessels; after the shape has cured, the form is removed from the inside of the vessel. A third method of manufacturing composites is to laminate alternating layers of resin containing structural fibers.

5.5.3 Ceramics and Glass

Industrial ceramics are made from the oxides of metals such as silicon, aluminum, and magnesium. Carbides, borides, nitrides, feldspar, and clay-based materials are also important ingredients. Ceramics are produced in many of the same ways as metals and plastics; extrusion, pressing, casting, injection molding, and sintering are all common methods. Most ceramic parts begin with a ceramic powder that may be mixed with other substances, depending on the required properties. Raw materials may be blended wet or dry with other ingredients, such as binders and lubricants.

Cold forming is the most common process in the ceramics industry, although hot forming processes are also used in some instances. Pressing techniques include dry pressing, isostatic pressing, and hot pressing. Slip casting is a common method of forming thin walled and complex shapes; this method is sometimes combined with the use of applied pressure or vacuum.

Extrusion is used to form continuous profiles and hollow shapes. This is done in a similar manner to that of plastic extrusion, although without the application of heat. The plastic form of ceramic material is simply the mixture of clay and water at ambient temperature. This mixture is forced through a die using a large screw called an auger.

Most ceramic materials must be heat treated after forming. This is necessary both to dry the ceramic form from its plastic state and to "fire" or harden the material into its final consistency. Intermediate processes such as sintering are also done to transform a porous form into a denser product by diffusing the material.

Final firing of ceramics is generally done in a furnace at a very high heat, often in the thousands of degrees. This causes the vitrification process to occur, where some of the components of the ceramic enter a glass phase, bonding some of the unmelted particles together and filling some of the pores in the material. This creates a very hard, dense, yet brittle material that may be used for many purposes.

Ceramics are used for insulators, in abrasives for grinding, as coatings on cutting tools, dielectrics for capacitors, for heat resistant containers, and many other products. The ceramic properties of hardness and resistance to high heat levels makes ceramics an important component in such parts as jet engine turbine components, engine valves, and thermal insulation tiles.

Glass is a substance that is made from inorganic materials, primarily SiO_2, or silica. It is manufactured by heating its ingredient to a liquid state or "fusion" and then cooled to a rigid state. Sheet glass is produced by a method called floating, where molten glass is floated on a bed of molten metal, usually tin. After cooling the sheet from about 1100°C to 600°C, the sheet can be lifted out of the bath and placed on rollers. The glass is further cooled as it passes through a kiln so that it anneals without strain. This produces a continuous ribbon of very flat and uniform glass, which is then cut into sections for shipment or further processing.

Laminating sheets of glass with a plastic interlayer in an autoclave produces shatterproof "safety" glass that is used for automotive windshields. Reheating glass to a semiplastic state and then cooling it quickly with air or "air quenching" produces tempered glass. This provides glass with greater mechanical strength and creates smaller, less dangerous pieces if broken. Tempered glass is often specified for windows and doors for strength and safety.

Glass containers are produced by pressing, blowing, or a combination of both. Bottles, jars, and lightbulb envelopes are formed by blowing molten glass into a mold. This process is similar to that described in the blow molding of plastics, where a partly manufactured container called a "parison" is reheated and blown into its final shape. This method is known as "blow and blow" and is used for narrow-necked containers. In the "press and blow" method, the parison is formed by a metal plunger pressing the solid piece of glass or "gob" into the mold. After the plunger is withdrawn, the parison is blown into the mold. A mechanism is then used to take the formed product out of the mold and the glass container is slowly cooled evenly, or annealed. Some containers undergo further treatment such as dealkalization—a chemical gas treatment—to improve the chemical resistance of the glass.

Fiber optics are formed in much the same manner as wire, by drawing a preform into a thin strand of glass. A hollow glass tube is placed horizontally on a lathe, where it is rotated slowly. The preform tip is heated, and optical fiber is pulled out as a string. Gases are injected along with oxygen as the heat is applied to optimize the fiber's optical properties. Strands are grouped into a fiber-optic bundle and then sheathed in plastic for durability and protection.

5.6 Assembly Machines

Individual components made from metals, plastics, and other elements must be combined to produce many of the consumer products in use today. Material handling, robotics, and many of the mechanisms and devices discussed previously are used to put these components together into a usable form.

Assembly machines may be fully automated or involve manual operations by a machine operator. Much of the assembly process involves the moving of individual product elements into proximity with each other and fastening them together. Many times a central processing path is used to move an item through an assembly machine or production line with peripheral processes feeding components in from the sides. Often a product is moved on pallets and indexed through a series of machines. Figure 5.8 shows an assembly machine for cables, parts are loaded by an operator.

5.6.1 Part Handling

Assemblies often start with one base piece, usually a larger component of some kind. Many products have a frame or housing of some sort that contains other elements. If an assembly processing line is pallet

Process Systems and Automated Machinery

FIGURE 5.8 Cable assembly machine.

based, this first element is placed onto the pallet either manually by an operator or with an automated device such as a robot or pick-and-place mechanism. Larger components may be presented in a palletized form or in a bin. Because of their size and often their method of shipping, these items may be difficult to handle with automation. Palletized parts are usually easier to load automatically since they can be precisely located within a formed pallet and the pallets can be made in a uniform and stackable size. Automated depalletizers are often built to handle these types of pallet stacks. Filled pallets enter on a conveyor, are destacked and emptied, and a stacker restacks the empties and sends them out on another conveyor. Stacks of pallets are generally handled by forklift.

Not all products can be easily handled by automation. Some basic components may be difficult or expensive to handle because of size or shape. Others may be hard to locate because of shipping and packaging methods. Items that are randomly oriented in bins or must be unwrapped may be more economical to load by hand.

Parts that are easily handled and located may be handled in a number of methods. If parts are presented in an array, such as within a formed or divided pallet or bin, they must be indexed, either by an actuator that can move a gripper to multiple X and Y positions or by a system such as an indexing table. These typically take the form of a pair of servo actuators arranged in a perpendicular manner. If parts are arranged in layers, a servo-operated Z-axis may also be required, otherwise the Z or vertical movement may be a simple pneumatic cylinder with a gripper, magnet, or vacuum cups on the lower end. Robots are increasingly used for this purpose because of their flexibility.

Intermediate part-handling methods are often used in assembly machines also. Components may be placed onto a dial table or moving fixture so that some kind of work can be performed on the product. Tooling to accurately locate the part is often mounted to the dial or fixture so that a device, such as a screwdriver or adhesive applicator, can precisely approach a point on the part. This tooling is removable so that it can be easily replaced or modified.

Feeders and conveyors are standard components on many assembly machine and production lines. Feeders usually have queuing and orienting outfeed sections that help locate parts precisely for pickup. Conveyors in assembly machines are often indexing conveyors with fixed stopping points. These may be cleated belt or chain type conveyors or chains with pendants that fixture parts accurately.

Simple methods such as pushers, lifts, pick-and-places, and guides are often used within assembly systems to move and maintain control of parts. A rule of thumb for handling parts is to never lose control, orientation, or containment once you have gained it. Parts may be oriented on a conveyor end to end, be singulated by pushing them perpendicular to their flow, and lined up in magazines or collated into rows. Once they are contained, they should never be placed back into a random arrangement such as a bin unless being rejected or removed.

5.6.2 Fastening and Joining

An important part of the assembly process is attaching parts and components to each other. After components of a product have been brought into close proximity and have been located accurately, one or both components are picked up or guided into contact. Parts may be connected using fasteners, adhesives, or using various welding methods.

Fasteners are often delivered from fairly standardized systems. Vibratory feeders are a common method of singulating and orienting screws, bolts, and rivets. Generally, fasteners are fed into a tool, such as a nut or screwdriver, by blowing oriented fasteners up a tube and into an intermediate escapement or buffering section. Queued fasteners are then brought individually to the tip or receiver of the driver and the fastener is installed either by an operator or automatically. Threaded or tapped holes are generally formed into the base workpiece, usually the larger of the two items being connected.

Most fastener systems are manufactured by OEMs who specialize in this type of equipment. A controller is generally attached to the spindle remotely. The drive mechanism is usually electrical so that torque can be accurately sensed and controlled. Controllers are easily integrated into control systems with digital I/O and communications interfacing to the main control system. Torque and screwdrivers almost always need to collect torque and angle (number of rotations) information for validation. Fastener systems are often operated manually because of the possible inconsistency of hole locations and angles. There is always the possibility of cross threading and the necessity of reversing the screw or bolt and removing it. Fasteners also may have flaws that require intervention by an operator. Automated fastener systems can be a major cause of machine downtime in assembly systems both in the feed system and the insertion.

Adhesive applicators use cold, air cured, or hot-melt systems to glue parts together. Like fastener systems, adhesive applicators and systems are usually made by OEMs with expertise in this field. Glues and adhesives are usually pumped from a reservoir to an application head. Cold adhesive systems are closed systems that prevent exposure of the adhesive to air until the glue is applied. For cold gluing applications, the glue may be applied using contact or noncontact methods. The distribution for cold systems may use a centralized tank with pipes to application heads or open-wheel pots, bulk containers with gravity feed or pumps through a siphon, or air-operated piston or diaphragm pumps. Pressurized tanks are another simple way to distribute adhesives.

Cold adhesives are generally acidic. Stainless steel piping is generally the best choice for the distribution system. Fiber-reinforced engineered plastics and PVC piping are sometimes used but must be protected from temperature extremes and physical abuse. After adhesive has been distributed to a point near the applicator, plastic tubing and noncorrosive synthetic materials are used to bring the adhesive to the applicator. Pressure regulators are usually placed between the pumping system and dispensing guns. Manual regulators can be used if application time and/or line speed are constant; automatic regulators may be required if pressures need to vary by product.

Applicators may be pneumatically or electrically operated. A tapered needle with a ball in a nozzle seat is moved to allow adhesive to flow. Electric guns integrate the solenoid into the applicator with the needle body located inside the solenoid coil. This allows the gun to respond more quickly than a pneumatic valve, allowing more control of the amount applied. The dispensing gun may spray the glue through a nozzle or pattern plate, apply a bead, or extrude adhesive through a slot. Slot dies are used to provide an even coverage and control edge position. They are often used in moving web applications. Spray guns and bead extrusion are both noncontact methods of application where the nozzle does not contact the part or substrate. Pattern plate and slot dies are both contact methods of application.

Hot-melt systems use a melter to bring a thermoplastic adhesive to a liquid state. Adhesives are loaded into a reservoir in pellet, block, or slat form and heated. A pump is then used to carry the adhesive through a heated hose to the valve, gun, or manifold, commonly known as the adhesive dispenser. Hot-melt systems are commonly used for packaging applications such as carton or case sealing and tray forming.

Both cold adhesive and hot-melt systems may be used in an automated or manual process. Automated application of adhesive may also use heads that apply small dots of adhesive before pressing parts together. Adhesive systems and applicators must be cleaned often and require periodic maintenance.

Many assembly machines and production lines use a combination of automated and manual processes for joining parts. The decision on which method to use may be based on cost or the difficulty of accessing and handling certain products. Robots may be used in some adhesive applications but are usually not cost-effective.

Other methods of joining components include ultrasonic welding of plastic components and welding of metals. Ultrasonic welding uses high-frequency vibrations to melt plastics together. Parts are placed in a fixed nest called the anvil and an acoustic vibration is transmitted through a metal horn, or sonotrode. Vibrations are generated using a piezoelectric transducer known as an ultrasonic stack. The stack contains a converter that converts an electrical signal into mechanical vibration, a booster that modifies the amplitude of the vibrations, and the horn that applies the vibrations to the part. The stack is tuned to resonate at a specific frequency, usually 20 to 40 kHz. The two parts are pressed together and an ultrasonic generator delivers a high-powered AC signal to the stack for an appropriate period of time.

Metal welding is generally done robotically in assembly machines. Figure 5.9 shows robotic welding of a metal tube. There are several methods of welding, but the most commonly used in automated applications are metal inert gas (MIG) or wire welding and laser or electron beam welding. MIG welding uses a wire feed as an electrode. An inert or semi-inert gas mixture is used to protect the weld from oxidization and contamination. The gas also acts as a shield to prevent porosity in the weld; porosity reduces the strength of the weld and may cause a pressurized vessel to leak. This method is sometimes called gas metal arc welding (GMAW). A related process called flux-cored arc welding (FCAW) uses a steel wire surrounding a powder fill material. This wire is more expensive than solid wire but allows for higher speed and penetrates deeper into the metal being welded.

Shielded metal arc welding (SMAW) is also called stick welding. It is also sometimes called manual metal arc welding (MMA) since it

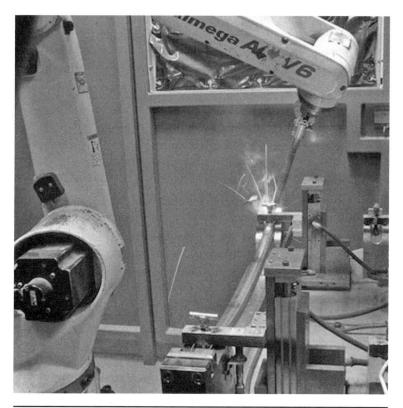

FIGURE 5.9   Robotic welding. (*Courtesy of Mills Products.*)

is not used in automated applications. Electric current is passed between a metal rod coated with flux and the workpiece. The flux produces a CO_2 gas that protects the weld area from oxidation. The metal rod acts as an electrode and is consumed during the operation. Stick welding is often used when deeper weld penetration is required than wire welding.

Plasma arc welding uses a tungsten electrode and a plasma gas to make an electrical arc. It is often used when welding stainless steel. A variation of this is plasma cutting, which uses air to blow the melted steel out of the workpiece separating it.

5.6.3   Other Assembly Operations

Many of the operations and processes discussed in the previous chapters are also used during the assembly operation. Converting operations, such as pressing or punching, are often performed as long as the residue material can be easily removed. Lubrication of mechanical assemblies and mechanisms is also often done before parts are sealed.

Machining and chip-producing processes are usually performed on parts before they begin the assembly process, but there are exceptions. Inspection of parts following critical processes is also common. This may be a simple part presence check using discrete sensors or separate systems, such as leak testers or machine vision. Adhesive application is often checked using ultraviolet sensors that look for dyes mixed into the adhesive. Dimensional checks are performed with LVDTs or other analog devices. Check weighing may be done on parts that are in transit or in stations where a part is set on a scale. Failed parts are either removed from the line as rejects or marked/identified for later removal. Parts that are indexed through the line can be tracked either through shift registers in the software or through tracking devices such as bar codes or RFID tags.

Marking and labeling are also common procedures in the assembly process. Adhesive labels, direct printing, and pin stamping are methods often used to imprint alphanumeric characters or bar codes onto products before being packaged.

5.7 Inspection and Test Machines

Inspection and gauging machinery is often built as a stand-alone unit separate from the assembly process. After completion of an assembly process, the final product may be inspected for quality, tested for function or gauged dimensionally. Many of the individual components of an assembly may have been checked during the assembly process also, but final testing is common for many complex products.

5.7.1 Gauging and Measurement

There are various ways of checking the physical dimensions of parts. Tooling can be used to ensure that parts fit where they are designed to. An example is placing posts in a fixture where bolt holes are located on a part. Discrete sensors such as photo-eyes and proximity switches can be used to detect the absence, presence, or location of features on an assembled part. Some manual assembly stations are outfitted with these types of "Pokayoke" checks built in.

Figure 5.10 shows a manual 6 LVDT gauging station for automotive parts. LVDTs often have pneumatic actuators that can extend to measure the position of parts relative to the fixture or to each other. This is not always desirable since tooling must contact the part, so machine vision and optical sensors are often preferred.

5.7.2 Leak and Flow Testing

Products often need to be tested to ensure that they do not leak or that airflow is within prescribed limits. Seals, filters, and gaskets may be part of the assembly being tested. In addition to ensuring their presence during the assembly process, leak or flow testing ensures that they are installed and are operating correctly.

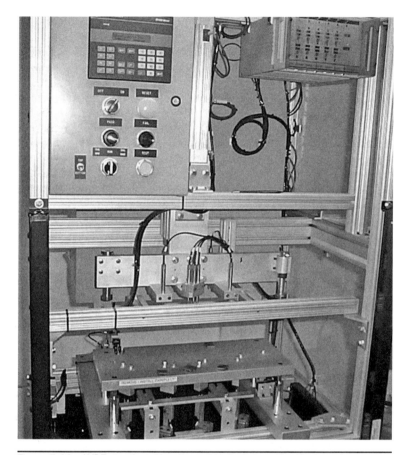

FIGURE 5.10 LVDT gauging. (*Courtesy of Nalle Automation Systems.*)

Systems may be as simple as a few valves, some pipe or hose, and a simple pressure transducer. If the purpose is simply to detect gross leaks, detailed validation procedures and calibration to a standard may not be required.

When performing tests that require more precision and a validation procedure, it is usually necessary to purchase a packaged system. As with many of the other special-purpose systems discussed here, these systems are usually manufactured by companies that specialize in the field rather than custom machine builders.

Purchased systems such as this have a lot in common, regardless of the manufacturer. The system is plumbed with various valves into an analog pressure transducer. There is an onboard controller with I/O connections for triggering and results, generally 24VDC. There are various communication ports for exchanging information with a PLC or computer as well as a separate programming port and usually a printer port also. These ports may be assignable as to communication protocols. Ethernet/IP, RS232, or open protocols such as DeviceNet or Profibus are common. The RS232 or Ethernet ports generally transmit a configurable string containing test data, time/date, and selected program information. This allows the parsing of the string(s) for the data that the user requires for archiving and display.

Fixturing the tested device usually involves some kind of rubber seal. This may be a solid surface that the product is pressed against or an inflatable bladder that is inserted into a round hole. Regardless of the type of sealing, the process usually involves movement of the part or tooling (or both), which brings the aspect of safety, that is, light curtains or physical guarding, into the picture. Figure 5.11 shows a manually loaded leak tester for automotive parts.

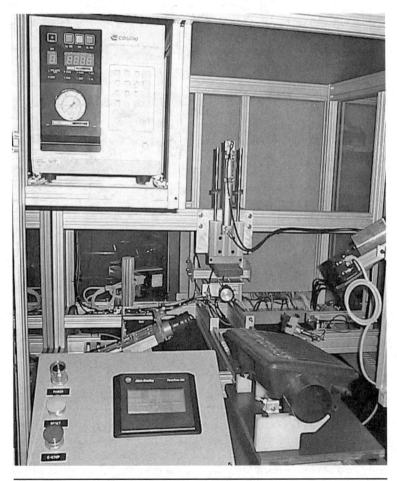

FIGURE 5.11 Leak tester. (*Courtesy of Nalle Automation Systems.*)

The interface generally allows the test parameters to be set by program number. Fill time, stabilize time, test time, and pass or fail criteria are settable within the interface. Alternatively, these can be externally controlled by means of I/O. A special mode for calibration allows the user to place a calibrated orifice into the flow circuit to test to a known standard. This can be automated for a periodic check.

5.7.3 Other Testing Methods

Products are often functionally tested by moving components and checking torque values, measuring electrical characteristics such as resistance or current flow, or applying power to a device and operating it. Many of these operations are often done at the end of an assembly line within a system, but "Test Benches" are often designed as a standalone unit to perform these checks.

Solid items are sometimes examined using ultrasonic testers or X-ray testers to locate internal cracks or flaws. Ultrasonic testing is often done in a water bath, so pumps and plumbing as well as material handling may be part of the machine or system. Safety and guarding are also important to ensure personnel are not exposed to harmful effects.

There are many different kinds of specialty devices, such as eddy current testers, which are used to check threading on the inside of tapped holes; ion source leak testing, which determines the size of tiny holes in materials; and various material testing systems for hardness or chemical composition. Some tests may be destructive, rendering the product unusable after testing. These types of test are usually done on a sampling basis.

CHAPTER 6
Software

Software is used to drive the motions of an automated machine or line and to gather data about its operation. It is also used to design and document automated systems. As with hardware, there are a wide variety of vendors for automation software. Manufacturers of control hardware usually also provide the software for programming their systems. This can often be quite expensive since the hardware manufacturer is usually the only provider for such software.

Design software may be proprietary and specific to machinery or a general-purpose package used for other applications. CAD and other visualization software packages, discussed in Chap. 2, are commonly used for the physical design of machinery and controls.

Analysis software is used to determine whether a solution will work before actually implementing it or to mathematically calculate parameters for physical systems. This software is sometimes combined with design software, such as three-dimensional modeling, or provided as an accessory to products, such as servo systems or robots.

6.1 Programming Software

Controllers, OITs, computers, motion systems, robots, and all manner of other complex automation devices make use of programming software in order to configure a target device to behave in a certain way. The programming languages vary from standard programming languages, such as BASIC, C, or ladder logic, to customized vendor-specific languages. Even the standard languages have wide variations within different vendor packages. PLC programming packages may all use ladder logic as a basis, but each manufacturer has developed their own interface and compilation strategy.

Operator interface software is typically graphical in nature. Buttons, set values, and indicators are drawn on a screen using standard tools, such as a mouse and keyboard. These items are assigned bit or word addresses in a controller for input or feedback. Pop-up text messages may be triggered from a controller to display machine status, and working parts of a machine or system may be animated to simulate the movement or status of actuators. Operator interfaces can be quite elaborate and are only limited by the imagination of the programmer and the information that may be exchanged with the controller.

Robot programming software usually takes the form of classic programming languages of the BASIC or Fortran variety. There are usually various modules or code sections associated with items, such as points, move trajectories, logic, and local I/O. Some of the programming is accomplished in combination with a "teach pendant" attached to the robot controller. This allows the robot to be moved to a point by controlling the separate axes or using real-world coordinates. The position may then be "taught" or entered into the robot's coordinate point list.

Individual or multiple servo axes are often programmed independently of a machine's master controller. This is because servos must react quickly to outside effects, such as an axis not being at the point where it is commanded to go. Movement of servos is monitored constantly by the axis controller to ensure that there is no deviation from the projected move. The programming of servos is usually done with the particular manufacturer's software and usually consists of configuration settings for preprogrammed moves and speeds, fault reactions, and other motion parameters. Servo controllers are usually tied closely with a master controller. Controllers often have specialized motion cards in their I/O racks to perform local motion control. These can sometimes be configured or programmed by connecting to the main controller.

Programs must usually be compiled before downloading them to a controller. Online or real-time programming can often be done, but the program is still usually converted to a lower-level "machine language" that operates more efficiently than its graphical and user-friendly counterpart.

6.1.1 Programming Concepts

Programming involves the entire process of design, coding, debugging, and maintaining software. A computer program is a series of instructions that tells a computer or microprocessor to perform certain tasks. In the automation field, the processor of these instructions may be a controller, such as a PLC, DCS, embedded processor, or robot controller. A computer may also execute instructions for data acquisition and display, database operations, or control functions. Programs may be written in a low-level, machine-specific language or in a high-level machine-independent language.

In automation, inputs and outputs are represented by digital values of "On" and "Off," True or False, or 1 and 0. Conditional statements such as "If," "Then," "Else," "And," and "Or" are used in combination with *variables* to make statements that create the desired result. Variables are names that may represent physical I/O, data values in string, number or boolean form, or internal data. Variables may be as simple as a single letter, as in the statement "IF X AND Y, THEN Z," or they can be more descriptive, such as in "IF Motor_On AND Button_Pressed, THEN Conveyor_On." Descriptive variables, such as those stated in the previous sentence, are known as *tags*. In a dedicated controller, such as a PLC or DCS, there are usually registers dedicated to different types of data. Tags may then represent different types of data, such as bits, integers, floating point or REAL, timers, and counters. Some PLC platforms allow the programmer to choose which registers are used for which data type and size memory accordingly.

The representation of a program may be text based, using alphanumeric characters, or graphical in nature. Ladder logic, used in PLC programming, uses representations of electrical components, such as contacts and coils. Some software uses blocks similar to flowcharts to program. The code inside these blocks may take another form, such as text or ladder logic. HMI and other visual programs have a graphical representation of the screen that will be displayed to the user. Objects on that screen are configured by entering the tag or variable that it interfaces with. Logical functions are sometimes embedded in the objects to execute when they are activated by the user.

Regardless of the software development platform, the final program must satisfy some basic requirements. The program must be *reliable*. Algorithms must produce the correct results; resources, such as buffer and data allocation size; should be correctly used; and there should be no logic errors. The program should also be *robust*; the program should anticipate problems not caused by programmer error, such as incorrect input and corrupted data. The program should be *usable*; textual, graphical, and even hardware elements should improve the clarity, intuitiveness, and completeness of the user interface. If possible, the program should be *portable*; software should be able to run on all the hardware and operating systems for which it is designed. Some software development programs only operate for the manufacturer's proprietary hardware. The program should be *maintainable*; present or future developers should be able to easily modify the code for improvements or customizations, fix bugs, or adapt it to new environments. Good practices and documentation during initial development help in this regard. Finally, a program should be *efficient*; the code should consume as few resources as possible. Memory leaks should be eliminated and unused code deleted.

6.1.2 Programming Methodologies

The first step in most formal software development processes is to analyze the requirements for the program. The scope of the requirements often determines whether a program will be written by a single programmer or by a team. The analysis phase involves reviewing deliverables and determining input, processing, output, and data components. Many programmers or teams will use what is known as an Input-Process-Output (IPO) chart to list these. If it is possible to meet with users of the software during this phase, it can be advantageous. This initial phase allows programmers to be sure they completely understand the scope and purpose of the software.

After ensuring that all the requirements of the program are known, the next step is to design the solution. This is a graphical or written description of the step-by-step procedures to solve the problem. The method that is chosen to design the solution will depend on the type of software platform and the preference of the programmer. There are two common methods used for solution design: process modeling, often called structured design, and object modeling, or object-oriented design.

In structured design, the programmer usually begins with a general design and moves toward a more detailed design. This method is sometimes called top-down design. The first step in top-down design is to identify the major function of the program, sometimes called the main routine or main module. This module is then broken down into smaller sections, called subroutines or modules. To document this procedure, programmers use a hierarchy chart or structure chart to show program modules graphically. This process is also known as flowcharting; Fig. 6.1 provides an example of this.

Structured design is simple, readable, and easily maintained, but it does not provide a way to keep the data and the program together. To eliminate this problem, some programmers use the object-oriented approach for solution design.

In object-oriented design, programmers package the data and procedure into a single unit, called an object. This concept of combining the data and program together is called encapsulation. By encapsulating an object, the programmer is hiding its details. The object sends and receives messages and also contains code and data. As an example, a print button or object is used to interface with a printer port and the attached hardware. When users access the object, they do not need to know how the procedure operates; they just use it. Programmers do need to understand how the object works so that they can use it effectively.

Objects are often grouped by type into classes. In standard computer programming, a designer uses a class diagram to represent classes and their relationships graphically. When programming a PLC or HMI, this may take the form of diagramming functions, special-purpose reusable code, such as servos, or OEM device interfaces. After completing a high-level class or function diagram, a designer should develop a detailed diagram for each class that provides a visual representation of each object, its attributes, and its

Software 223

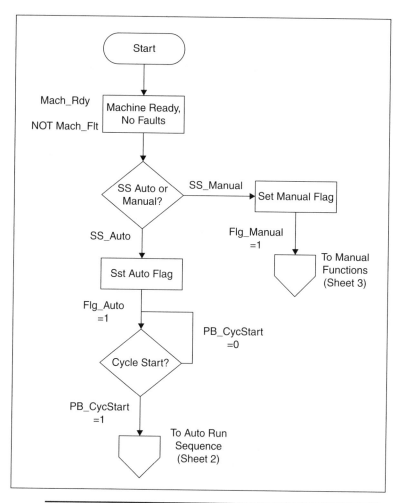

FIGURE 6.1 Structure or flowchart.

methods. The methods are then translated into program instructions that perform the required tasks.

After the solution is designed, the third step is to validate the design. This step consists of reviewing the code for accuracy and, if possible, testing it with real data. There are several methods of validation, the simplest is known as desk checking. In this method, various sets of test data are developed and expected outputs determined independent of the algorithms. The code is stepped through using this data, ideally by a different programmer, and results are recorded. Expected results are then compared with actual results and faulty code is corrected. Another method is that of emulation. Many software packages allow for the operation of a virtual machine or emulator. For physical machinery, inputs can be manipulated, HMIs can be operated, and code simulating processes can be run. This code is often time based; as an example, for a synchronous machine, a set of variables can be manipulated simulating the indexing of a chassis, and the relationship between stations can be monitored.

Implementation of the design is the fourth step in the software development process. If the program is simply accessing and manipulating data, there are usually no safety issues concerned other than security and the possible interruption of processes running on other machines. Implementation may simply consist of installing software, compiling the program or programs, and downloading them. In the case of software that operates moving machinery, I/O and electrical checks, safety system testing and hazard reviews with personnel must be completed before any code is activated. When using a standard personal computer, the compiled program is kept on the hard drive. It is loaded into random-access memory (RAM), along with variables and their values, when the program is started. A database is often used to store these variables and their values when power is removed from the system.

After implementing the program, it must be tested and debugged. This is to ensure that the program runs correctly and is error-free. In general, debugging of computer code involves finding two types of error: syntax errors and logic errors. Syntax errors are often flagged by the programming software using a debug utility. Incorrectly spelled commands, duplicate outputs in PLC programs, or illegal operations are examples of syntax errors. Logic errors often take more time to locate and cannot be done automatically. They often involve the movement of actuators on machines and may only be detected when an operation malfunctions. Debugging can be time-consuming; code needs to be tested by operating software or machinery correctly and by introducing incorrect data or purposeful fault conditions. Since many conditions are not easy to anticipate, sufficient time must be allocated to try many different kinds of unusual circumstances.

Debugging of automation machinery involves operating actuators and moving parts under varying conditions and observing results. As such, it is an iterative process; often when solving one issue it will reveal another. Introducing anomalous conditions—such as E-Stops, flagging sensors at inopportune times, or introducing or removing parts—may be part of the debugging procedure.

The final step in the development process is to document the program and clean up the code. Most programmers comment on their program as they write code, but it is always a good idea to have peers examine the program to ensure that it is easy to read and well organized.

6.1.3 Languages

Programming languages are chosen based on several different factors. Companies may have a standard that they use a certain programming language because that is what their employees are familiar with. Customers may have a standard with which they require all vendors to comply. Many hardware platforms are only programmable with software sold by that vendor. Some software packages may be easier to learn than others or have capabilities not available in other packages. Of course, cost must also be considered. Ideally, the programming language best suited for the task at hand will be selected.

Certain instructions and functions appear in any programming language. Programs must be able to handle *input*. This may be from a computer keyboard, an HMI, sensors attached to an input device, or from a file. They must be able to deliver *output*—displaying data to a screen, turning on physical outputs on a device, or writing data to a file. Instructions must be able to perform *arithmetic*, whether simple addition and multiplication or complex trigonometric and interpolation functions. They must be able to execute *conditional* statements, making decisions based on information and performing the appropriate functions in response. Finally, they must be able to *repeat* actions, often with some kind of variation.

Many computer languages also provide a mechanism to call functions provided by libraries. Provided the functions in a library follow the same method of passing arguments, then these functions may be written in another language. The exception is with closed architectures, such as PLC programming software. Most software packages, even proprietary ones, do, however, allow data exchange with such protocols as object linking and embedding (OLE), OLE for process control (OPC), and open database connectivity (ODBC). ActiveX component interfacing is also common.

Computer Languages

Source code, the written instructions that make up a program, cannot be directly understood by a computer or microprocessor. There are many different languages, but all must be converted or "compiled" into a hexadecimal machine language that a processor can understand. Machine language or machine code instructions use a series of binary digits arranged in groups of hexadecimal numbers. These are not visibly interpretable by a programmer since they do not use meaningful instructions or codes. Machine language is also known as first-generation code.

The second generation of code is a low-level language known as assembly language. This is a language written in symbolic instruction codes that are meaningful abbreviations, such as A for Add, C for Compare, LD for Load, and so on. While easier to write and understand, coding in assembly language is very tedious and not

Start:	.org$6050		Code Starts at Line 6050
Address	Assembly	Hex	Comment
6050	SEI	78	Set Interrupt Disable Bit
6051	LDA #$80	A9 80	Load Accumulator HEX 80 (128 Decimal)
6053	STA $0315	8D 15 03	Store Accumulator to Address 03 15
6056	LDA #$2D	A9 2D	Load Accumulator HEX 2D (45 Decimal)
6058	STA $0314	8D 14 03	Store Accumulator to Address 03 14
605B	CLI	58	Clear Interrupt Disable Bit
605C	RTS	60	Return from Subroutine
605D	INC $D020	EE 20 D0	Increment Memory Address D0 20
6060	JMP $EA31	4C 31 EA	Jump to Memory Address EA 31

FIGURE 6.2 Example of assembly and hexadecimal machine languages.

efficient for the programmer. Figure 6.2 shows a section of assembly language code with its hex or machine code equivalent.

Procedural languages, often known as third-generation languages, use English-language words like ADD, PRINT, IF, and ELSE to write instructions. DO WHILE and FOR NEXT are common program control and repetitive task instructions. They also use arithmetic operators such as "*" for multiplication and "/" for division. Third-generation languages must be interpreted or compiled into machine language also. The resulting machine language program is known as object code. Compilers convert the entire program before executing it, while interpreters translate and execute the code one line at a time. An advantage of interpreted code is that if an error is found, the interpreter will provide feedback immediately. Many languages have both an interpreter and a compiler for easy program development. COBOL and C are examples of third-generation languages.

Other "classic" programming languages include BASIC, Fortran, and several others. Variations of these are often used in OEM equipment to perform logic functions such as printing and presenting test or operation results.

Object-oriented programming (OOP) uses items that can contain both data and the procedures that read or manipulate it. These items are called objects. An advantage of OOP is that once an object has been created, it can be reused and modified by other existing or future programs. An object represents a transaction, event, or physical object. In addition to being able to work with objects, OOP languages are event driven. Events are actions, such as keyboard or other device inputs. These may be generated by pressing a button on an HMI, a sensor being activated in a control system, or typing a value into a text box. Examples of OOP languages are Java, C++, C#, and Visual Basic.

PLC Languages

The IEC has defined an open international standard for PLCs. IEC 61131-3 is the third section of eight and deals with programming languages. Three graphical and two textual languages are defined in the standard. Text-based languages are instruction list (IL) and structured text (ST), while the graphical languages are ladder diagram (LD), function block diagram (FBD), and sequential function chart (SFC).

The standard also defines data types and variables, configuration of PLCs, program organization units, resources, and tasks. The elements in this standard had already existed for many years before the standard was created in 1993, but the definition helped to pull many of the diverse elements of different platforms into a common framework.

Data types in PLC programming are defined as follows:

Bit strings (groups of on/off values):
    BOOL—1 bit
    BYTE—8 bit
    WORD—16 bit
    DWORD—32 bit
    LWORD—64 bit

Integers (whole numbers):
    SINT—Signed Short, 1 byte
    INT—Signed Integer, 2 byte
    DINT—Double Integer, 4 byte
    LINT—Long Integer, 8 byte
    U—Unsigned Integer (prepend to type to make unsigned)

Real (floating point IEC 559, IEEE):
    REAL—4 byte
    LREAL—8 byte

TIME (duration for timers, processes)

Date and time of day:
    DATE—(calendar date)
    TIME_OF_DAY—(clock time)
    DATE_AND_TIME—(time and date)

STRING (character strings surrounded by single quotes)

WSTRING (multibyte strings)

ARRAY (multiple bytes stored in the same variable)

Subranges (limits on values, such as 4 to 20 current)

Derived (derived from the above types):
    TYPE—Single Type
    STRUCT—Composite of several variables and types (that is, UDT)

Data elements may be held in predefined registers, or, in some cases, their location may be completely assignable and configurable by the user.

Variables are defined within the standard as being global (accessible by all routines), direct or local (accessible only by the containing routine), I/O mapping (inputs and outputs), external (passed from an outside source), and temporary. Configurations of the program are also defined. Resources are items originating from or pertaining to the CPU itself. Tasks are groups of programs or subroutines of which one is usually designated as the "main." There may be multiple mains per CPU. Programs and subroutines then run under tasks and form organizational units or subunits.

Other program organization units include functions, such as ADD, SQRT, SIN, COS, GT, MIN, MAX, AND, OR, and others, that may also be custom or user defined. Function blocks are containers for these functions. They may be customized or can be available as libraries from vendors or third parties.

An LD, or ladder logic, is a graphical representation of physical coils and contacts derived from the time when relays were used to control systems. Program elements are arranged in horizontal lines called rungs that simulate an electrical circuit. These lines are drawn between two vertical lines called rails. Contacts, coils, timers and counters, and various data operations are arranged along the rungs of the diagram. The resulting graphic looks very similar to a ladder, hence the name. Figure 6.3 shows a section of ladder logic.

Contacts may be NO, NC, or various forms of one-shot or single-scan contacts. Coils are similar to coils of a relay and may be active only if the preceding logic is true or may retain its state as a latch-unlatch or set-reset pair. Constructs such as timers, counters, and mathematical functions are predefined as elements in ladder software. Some PLC software manufacturers have hundreds of predefined special-purpose functions, while others allow the creation of functions by the programmer.

Rather than operating linearly and waiting for an instruction to step before the program proceeds, a PLC *scans* the entire program from start to finish, then updates the internal I/O table many times per second. This is one reason why PLCs are considered more deterministic than computers. It can also create race conditions if a programmer is not careful. This is where operations do not happen in the order intended, because rungs are active elsewhere in the program. This condition is not always dependent on where the operation is located in the program.

Because of the inherent program difficulties associated with scanning, *sequencers* are often used to control program flow. These define the state of an action or process by numerical "equal" statements or bit-of-word logic. A common use for a sequencer would be in the definition of a step-by-step automatic sequence to control a mechanism such as a pick-and-place.

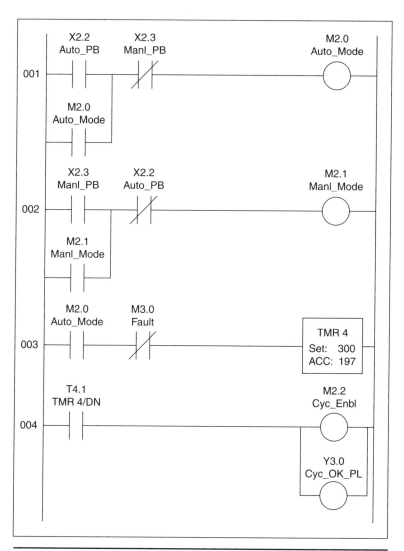

FIGURE 6.3 Ladder logic.

Because PLC programming software is developed by different companies for different hardware, the symbols or names of variables, devices, and techniques can vary widely. Inputs are commonly labeled I or X, while outputs are usually O, Y, or Q. Timers are usually T and counters C, while internal bits and variables can be nearly anything. Techniques for performing mathematical functions may use a single block with multiple variables, such as ADD, with A and B being added to produce result C. Another technique is to use separate commands, such as Load A (LD A), Load B (LD B), ADD,

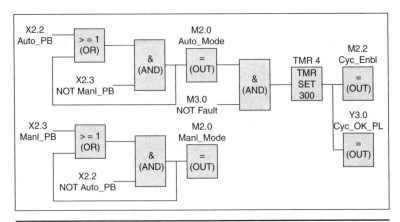

FIGURE 6.4   FBD.

OUT C. These would typically be placed in the same rung; the load command is loading a register known as the accumulator.

FBDs use boxes and lines to indicate program flow. The diagram describes functions between input and output variables, using arrows to connect blocks signifying the movement of data. Single lines called "links" are used to connect logical points within the diagram. The lines are oriented with arrows at one end. Figure 6.4 shows an example of FBD programming.

Links are drawn to connect an input variable to the input of a block, the output of a block to the input of another block, or the output of a block to an output variable. Multiple right side connections can also be used with a junction point, these are known as divergences.

SFCs are another graphical method of programming. SFC programming is based on a graphical language called Grafcet, which is a method of representing automated systems and logic flow. The components of SFC are *steps*, along with their associated actions; *transitions* with their logic conditions; and directional *links* between steps and transitions.

Steps in SFC may be either active or inactive. They are activated either upon being designated as an initial step by the programmer or by preceding logic. If all the steps before a step have activated and the logic connecting them becomes true, then the current step will become active. Variables associated with a step may be Set (S), Reset (R), or Continuous (N). Set and reset commands are latched and unlatched, while the continuous action N is only on if the step is active. Figure 6.5 is an example of an SFC.

A sequence of steps in SFC is abbreviated as a POU. Multiple POUs can be active at once, making SFC a parallel language. Outputs and variables from one POU can be used in another, an action known as "forcing." LDs can also be used within the blocks of an SFC

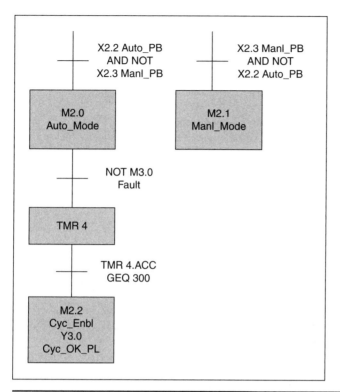

FIGURE 6.5 SFC.

diagram. Since SFC was derived from a graphical design tool, it is very easy to test, maintain, troubleshoot, and design with.

IL is a low-level text-based language similar to assembly language. Prior to the use of graphic terminals and PCs, ladder logic had to be entered into the PLC using a handheld terminal with a keypad. The language consists of many lines of code with each line representing one operation. Mnemonic codes are used for operations and addresses must be referred to directly, that is, no symbols or comments.

Instructions and expressions use a memory construct known as a *stack*. Instruction values are entered line by line and "pushed" onto the stack. After all lines have been entered for a rung, the logical calculations within the stack are performed up to the output device statement. Figure 6.6 shows an example of IL programming.

IL programming is the most fundamental level of PLC programming languages; all ladder programs can be converted to IL. IL programs cannot always be converted to ladder, however, since it is easily possible to build illegal rungs. If instructions compatible with IEC 61131 are used, IL can be used to migrate programs from one vendor platform to another.

Figure 6.6
IL.

```
LD X2.2 Auto_PB
O M2.0 Auto_Mode
AN X2.3 Manl_PB
= M2.0 Auto_Mode
LD X2.3 Manl_PB
O M2.1 Manl_Mode
AN X2.2 Auto_PB
= M2.1 Manl_Mode
LD M2.0 Auto_Mode
AN M3.0 Fault
= TMR 4 Set 300
LD T4.1 TMR 4/DN
= M2.2 Cyc_Enbl
= Y3.0 Cyc_OK_PL
```

ST is a high-level textual language based on PASCAL. Variables and function calls are defined by IEC 61131; they share common elements with other languages within the standard. Because of the adherence to the standard, a ladder logic program can call a structured text subroutine.

ST programs are composed of statements separated by semicolons. Programs start with statements defining variables, and the program begins following these declarations. Programs then use predefined statements, subroutines, and variables to execute the code. Good practices, such as indenting and comments, should be used as with any programming protocol. ST is not case sensitive, but making statements uppercase and variables lowercase can be a useful technique to enhance readability. Figure 6.7 shows an example of ST programming.

Unlike linear programs used in computers, ST is scanned continuously as indicated in the ladder logic description above. Because of this difference it is important to ensure DO-WHILE or FOR-NEXT loops do not take too much time to execute. PLCs constantly monitor the time it takes to complete a full scan with a watchdog timer. If the configured time is exceeded, a controller fault will be created.

Organization of code in all the previously listed programming methods uses tasks, programs, and subroutines for program control and separation. Usually, a main routine is used to call the subroutines. A common method of organization is to place inputs, outputs, sequences, faults, and system/mode control in separate subroutines. If a machine or line is fairly large, these subroutines are then placed in groups of program "cells" or "stations" for organization. For example, a device like a dial table, pick-and-place, or robot may be placed in its own program with subroutines for inputs, outputs, and

Software

```
// PLC Configuration
CONFIGURATION DefaultCfg
VAR_GLOBAL
    Auto_PB     :IN @ %X2.2       // Auto Pushbutton
    Manl_PB     :IN @ %X2.3       // Manual Pushbutton
    Cyc_OK_PL   :OUT @ %Y3.0      // Cycle OK Pilot Light
    Auto_Mode   :BOOL @ M2.0      // Automatic Mode
    Manl_Mode   :BOOL @ M2.1      // Manual Mode
    Cyc_Enbl    :BOOL @ M2.2      // Cycle Enable
    Fault       :BOOL @ M3.0      // Machine Fault
    TMR 4       :TIMER @ T4       // 10ms Base Timer
END_VAR

END_CONFIGURATION

PROGRAM Main

STRT  IF (Auto_PB=1 OR Auto_Mode=1) AND Manl_PB=0 THEN Auto_Mode=1
      ELSE IF (Manl_PB=1 OR Manl_Mode=1) AND Auto_PB=0 THEN Manl_Mode=1
      End IF

      IF Auto_Mode=1 AND Fault=0  THEN
      START TMR 4
      END IF

      IF TMR 4.ACC GEQ 300 THEN
      Cyc_Enbl=1
      Cyc_OK_PL=1
      END IF

      JMP STRT

END_PROGRAM
```

FIGURE 6.7 ST.

so on. Tags within each program or cell are considered local and can only be seen by routines within the program. Tags scoped for the controller are global and can be seen by all programs. Figure 6.8 shows the organization of a sample program in Allen-Bradley's ControlLogix software.

Well-known PLC programming software packages include Allen-Bradley/Rockwell Software, Siemens, Modicon, Omron, Mitsubishi, Automation Direct/Koyo, and many more. Not all software makes use of the IEC 61131 standard for compatibility, but all are programmable in ladder and can usually also be viewed as an IL.

Graphical User Interface (GUI) Programming

Computers allow operators to interact with software by typing commands on a keyboard, clicking images with a pointing device, such as a mouse, or touching a touch screen. Interfacing with the

FIGURE 6.8
Code organization.

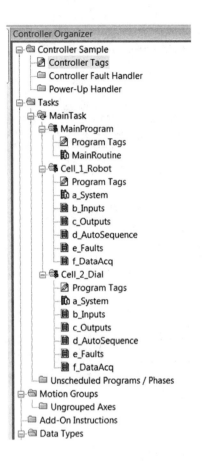

computer or a controller is made easier by the use of images and graphics to guide the user through operations.

The goal of any GUI is to allow the user to interface with the software and underlying system in a simple and intuitive way. Location of similar devices in the same position on each screen and laying out elements in the order in which they are used are both important techniques in interface design. The user should be able to navigate through the screens with little or no instruction. GUI design should be an iterative process with improvements being made constantly as the software is developed.

The Windows operating system is an example of a GUI. Commercial programs that run in Windows have familiar elements such as menus, radio buttons, icons, windows, and check boxes. Some programs also use buttons that reference the function keys on the keyboard. This technique is carried over into many operator interfaces both with and without touch screens.

Custom GUIs may be programmed in a variety of languages. Most languages have common devices such as buttons, text and numerical inputs, and the ability to interface with objects and components. Visual Basic was one of the early platforms used for GUI programming and is still used. Other common programming languages are Java, .NET, Python, C/C++/C#, and Perl, among others. Figure 6.9 is an example of a robot programming software GUI run on the Microsoft Windows operating system.

There are important questions to consider when developing a GUI for computer use. How easy is the language to learn and implement? Do you need customized controls, and will it interface with different objects? Will it need to run on different platforms, that is, Windows, Unix, Apple? For simple utilities running on a Windows platform, Visual Basic 6 is probably the easiest to learn. More powerful—but still relatively easy to learn—is C#, while Java is a good choice for running across different platforms.

Most of the larger PLC manufacturers also make GUI development software for the PC. They tend to cost more than the above listed programs, often being priced by the number of tags that will be used. These packages are generally easier to learn and have built-in drivers and configurations for PLCs. They also interface with objects such as ActiveX components and use OLE. OPC and ODBC are also standard. If developing a GUI for a specific brand of PLC, this is often a good choice.

In automation, if a computer is chosen over a simpler HMI it is usually because of requirements like data archiving and retrieval, interfacing with other computers on a network, and access to other software on the same computer. Computers are usually not as robust

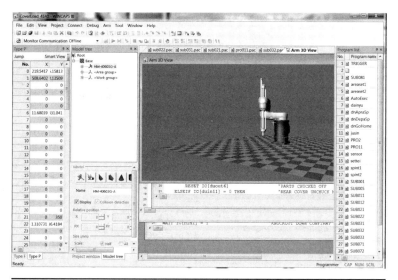

FIGURE 6.9 Windows GUI.

Chapter Six

for plant floor applications as dedicated HMIs and can be expensive when industrial versions are used. Because of this, hybrid HMIs running Windows CE are often used as HMIs. These can interface with a server to log data while still booting up faster than a standard PC.

For standard HMIs, the software development package is provided by the manufacturer. In most cases, they can be interfaced with several different PLCs, although functionality may be limited outside the manufacturer's product. PLC bits and words are addressed either directly (by address) or through the use of tags. Tags can often be imported directly from the PLC or through an intermediate step like importing a .csv (comma separated file) through Microsoft's Excel. Figure 6.10 shows a GUI developed for a machine HMI.

Automation-related GUI and HMI software generally has a number of features not found in open packages like Visual Basic and Java. Because the software interfaces with machinery, alarm display is an important part of the package. Alarms are created based on bit status or the numerical range of tags. They can be prioritized and displayed in order of priority. Buttons for acknowledging and clearing the alarm are standard objects. Alarms can be made different colors and sometimes have an alternate text field that can display a corrective action. A history display is often standard in a software package that allows a configurable number of alarms to be archived along with time of occurrence, time of acknowledgment, and time of reset or fault correction.

Graphics can be imported to assist the operator in identifying machine sections. Indicators are often superimposed over graphics to draw attention to an occurrence, such as a blinking red icon for an E-Stop or a red rectangle for an open guard door. Graphical

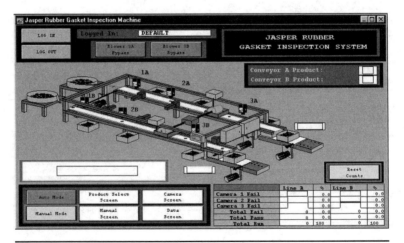

FIGURE **6.10** HMI GUI.

representations of actuators may be "animated" or moved on the screen to show their position. Sensor indicators are often placed on the screen to show actuator status.

Push buttons are configured with both a button tag and an indicator tag. If the button is used to control an actuator in manual mode, the button tag sends the command to the controller, while the indicator is used to indicate its position. The position status is generally directly indicated by the state of the sensor; although if a sensor is not used, it may be controlled by the output command status.

GUIs for automated machinery are often programmed in a hierarchical manner, with a main screen and various "trees" of displays. It is common for a graphic of the entire machine or production line to be placed on the main or "start-up" screen with areas of the graphic acting as screen select buttons leading to cells or stations. Other buttons placed on the main screen may lead to production data displays, calibration screens, fault histories, password setup, and I/O status screens. Security is usually a part of the GUI programming package, with different levels of security giving access to password-protected screens.

Robot Programming

Robots are programmed using a combination of a computer and a teach pendant. There are two basic types of data that must be programmed or taught: procedures and positional data. The setup and programming of motions and sequences is usually done by linking the robot controller to a computer with the programming software installed. Initial communication parameters are set up using the pendant to allow the controller to interface with both the programming computer and any other controllers.

Once the programming computer has been connected, programs and positional data can be transferred from computer to robot controller and vice versa. Program design for robots is similar to that for machines in general; it is usually a good idea to start with a flowchart or sequence of events. Articulated robots perform by storing a series of positions in memory and moving to them at various times in their programming sequence. For example, a robot that is installing a part on a workpiece might have a simple pick-and-place program similar to the following:

Define points P1–P5:

1. At home position (defined as P1)
2. 10 cm above bin with part (defined as P2)
3. At position to take part from bin (defined as P3)
4. 10 cm above workpiece (defined as P4)
5. At position to release part at workpiece (defined as P5)

Define program:

1. Move to P1
2. Move to P2
3. Move to P3
4. Close gripper
5. Move to P2
6. Move to P4
7. Move to P5
8. Open gripper
9. Move to P4
10. Move to P1 and finish

After defining points and events, the program is written using the computer software. Software usually uses procedural languages as described in the computer language section (Sec. 6.1.3). Decision-based structures, such as IF-THEN-ELSE and logical statements, are standard in robot software.

Programs may be much more complex than the example above. Handshakes using I/O from other controllers are common. These may be physically wired into points and addressed as local I/O or be interfaced using communications methods such as DeviceNet, Profibus, or Ethernet/IP.

A robot and a collection of machines or peripherals are often referred to as a cell or workcell. A typical cell might contain a parts feeder, an assembly station, and a robot. A PLC is often used to control the other stations and provide direction to the robot. How the robot interacts with other machines or stations in the cell must be programmed, both with regard to their positions in the cell and synchronizing with them. I/O mapping is a common element of a robot program, configuring inputs as commands to the robot and outputs as status bits. The cell controller may step the robot through every motion individually, or the robot may perform a sequence of moves, only stopping for interrupt or fault conditions.

Positions may be taught in several ways. X-Y-Z positions may be specified within the program, using a GUI or text commands. This technique has limited value because it relies on accurate measurement of the positions of the associated equipment and also relies on the positional accuracy of the robot. Robot positions can be taught via a teach pendant. Common features of such units are their ability to manually send the robot to a desired position or "inch" or "jog" to adjust a position. They also have a means to change the speed since a low speed is usually required for careful positioning or while test-running through a new or modified routine. This is probably the most common method of teaching positions on articulated robots. Another technique offered by many robot manufacturers is "Lead by the Nose." In this method, one user holds the robot's manipulator, while another person uses the pendant to deenergize the robot, causing it to go limp. The user then moves the robot by hand to the required positions and/or along a required path while the software logs these positions into memory. The program can later run the robot to these positions or along the taught path. This technique is popular for tasks such as paint spraying.

A common device interfaced in robot software is machine vision. There are basically two different ways to use machine vision with robots: mounting the camera on the robot's end effector tooling, so that wherever the tooling moves is what the camera is looking at, and mounting the camera in a fixed position to look at the robot's area of operation.

The first thing that has to be done is to place the two systems on the same coordinates. If the robot is at an angle to the camera's field of view (FOV), the robot's work space or frame reference must be changed. For this, a calibration grid is used, which is a printout of a checkerboard pattern with squares of a known dimension.

The printout is placed in the approximate center of the camera's FOV so that the x-axis of the pattern lines up with the camera pixels. The robot is then taught two points on the x-axis and entered into a "work" or frame variable. When this work space is invoked, the robot's X-Y space is then referenced to this variable.

The camera is then used to capture the image and save it. Cognex has an algorithm for calibration that places crosshairs at all the intersections on the grid. One set is then chosen as the origin, and the X and Y directions are defined from there. The robot is then used to locate the coordinates of the origin, and the data is entered into the camera. After the grid spacing is entered, the calibration algorithm is triggered, and the camera can then report the location of objects within the FOV in real-world robot X-Y coordinates.

Because objects at the edge of the FOV are farther from the lens, a parallax is created where coordinates have to be slightly scaled—this is another feature of the Cognex vision system. This is common in vision applications where measurement data must be very accurate. In some applications, the X and Y coordinates, as well as the rotation of the target object, are sent directly to the robot controller from the vision system. These values can then be massaged slightly to create a pickup point for the robot. The Z value will be a constant if picking from a one-dimensional array of parts. If not, another camera can be used to capture the position or an array of offsets can be used.

Using a camera with a fixed FOV is fairly straightforward because the coordinates remain constant, as does the focus. If the camera is mounted on the end effector tooling, the coordinate system needs to follow the robot's position and the focus may need to vary. This brings up an entirely different set of issues to be addressed. It is also common for the camera to be offset from the gripper tooling in both cartesian and rotational coordinates.

6.2 Design Software

Design software is often used to produce drawings in two or three dimensions of a physical system. 3-D modeling packages such as AutoCAD Inventor, SolidWorks, and Pro-E are used by a designer to lay out machines or manufactured products. Electrical and pneumatic circuits and other systems are often drawn in a 2-D drafting package such as AutoCAD or AutoSketch. These packages are often customized for the discipline of the designer. For instance, an electrical designer may have no need for three-dimensional drawings but may require libraries of special devices or cross-references to catalog part numbers. Many component manufacturers have made their hardware available in a variety of drawing formats for these libraries.

Design software is also available for devices such as CPLDs, FPGAs, ASICs, and other processors. These devices are used in embedded controllers, usually in OEM equipment. This software allows the user to configure electrical systems for board-level components, laying out architectures and power management and performing simulations. This software is also a useful tool for board-level vision systems and image processing. Altera is a manufacturer of both components and design software for system-on-a-programmable-chip (SOPC) design.

Special-purpose software for development of P&ID diagrams or other specialized documents contain tools allowing the developer or designer to quickly generate drawings for their purpose.

6.3 Analysis Software

Analysis software is used for sizing servomotors, determining stresses on mechanical systems and constructions, or calculating any factor in the design process. Engineers often create their own analysis software on platforms like Microsoft Excel, MATLAB, or LabVIEW.

Vendors often provide software free or at a nominal fee to assist purchasers of their products in selecting a proper solution. Allen-Bradley's Motion Analyzer software is a useful program for configuring servo systems. It allows the user to enter information about the application, such as size and physical attributes of the applied load, required speed, available power, and duty cycle. It then calculates inertia and other physical parameters and provides a selection of motor and gearbox pairs to satisfy the solution. Siemens Drive ES does the same. Because these are provided by the hardware vendors, the motor choices are limited to those provided by the vendor.

Pneumatic hardware vendors also provide sizing software for their products, allowing the user to provide weights and speeds along with available air for the application. Valve banks and actuators are then selected by the software—again, only from the vendor's products. Details such as cylinder style, mounting methods, and sensors are all configurable. Festo, SMC, and Numatics all have useful software for this purpose.

Pneumatic, hydraulic, and servo software is also available from third-party vendors; however, it mostly provides simulations and does not select product numbers for the user. Instead, it sizes components by kW, bore size, and so on, and leaves it to the user to choose components.

Factory physics and simulation software can be used to model a plant in 3-D. Routing of products, production times, shift calendars, and downtime can be programmed into the simulation and used to determine bottlenecks in a process. Since the simulation can be sped up and run using different product, process, and resource combinations, this type of software can help optimize performance and make production more efficient. Popular factory simulation software packages include AutoMod, eMPower, ProModel, FlexSim, the UGS VisFactory suite, including VisSim and VisProduction, and Adept Technology's Production Pilot software.

Machine simulation software can also be used to simulate product and actuator movement in a production line. Most 3-D CAD-CAM programs allow for a certain amount of simulation within the solid model to determine interferences and maximum rates. From a machine-building perspective, simulation software can help avoid risk and correct unbuildable conditions. SolidWorks, Pro-E, and AutoCAD Inventor all have the ability to simulate machine movement.

6.4   Office Software

Various office-type software products are used as part of the documentation process. Spreadsheets are used to document bills of material and capture costs, databases are used to archive data for easy retrieval, and word-processing software is used to generate written documents describing maintenance and operation of a machine or system.

Project management software allows schedules to be created and tracked as the implementation of a project progresses. Milestones are created and tasks can be shifted with dependencies being moved in concert. The number of resources required to finish a project by a certain date can be determined along with visually representing the critical path of a project. Estimates can then be adjusted and their impact viewed as a project progresses, creating a record that can be analyzed upon project completion.

Presentation software allows the visualization of a project or plan for meetings. Flowcharting software can be used for software design or project flow. Contact management and live meeting software allow people in different places to communicate with each other effectively.

Because of the widespread use of the Microsoft Windows operating system, most of the office software used in industry is compatible with the Microsoft Office software suite. This suite includes Word, Excel, PowerPoint, Visio, Access, Outlook, and several utility applications. Microsoft Project is most often used for project scheduling and tracking.

Macros within spreadsheet packages are also used extensively for data collection, calculations, and display. These are marketed by third-party developers to operate within a standard spreadsheet like Microsoft Excel. They allow users to turn data into histograms, Pareto charts, and other analysis graphs and displays.

6.5 SCADA and Data Acquisition

Supervisory control and data acquisition (SCADA) packages are used to both control automated systems and gather information about the processes. SCADA is nearly always installed on a computer or computers and is usually networked to other controllers on the plant floor. Controllers act as collection points for machine or node-specific data and often act as a backup for the collected information. Operator terminals may also be used as data entry points for operators, supervisors, or engineers. These may or may not be attached to a machine controller.

One of the major purposes of a SCADA system is to archive and share data. Data may be gathered on a periodic or event-driven basis. As an example, pressure values from various points may be logged on a periodic timescale, such as minutes or hours, or values may be recorded when a pressure crosses an alarm threshold. Values are often recorded to standard database platforms or formats such as .dbf files or SQL databases. This allows the data to be manipulated with third-party software packages for statistical analysis or archiving.

Computers running the SCADA software packages are connected to sensors through controller communication ports or through cards mounted in the computer itself. Some manufacturers have developed their own hybrid computer systems by placing a computer card in a chassis containing I/O and communications cards rather than using off-the-shelf computer systems. This can have the advantage of making for a more rugged system, but it can increase system cost and make it more difficult to find replacement components.

Data acquisition systems can often run without being connected to a computer. These are similar to SCADA systems except that they often run without having a graphical interface.

6.6 Databases and Database Programming

Automation systems must often save data in an organized and easily retrievable way. Production data—such as machine OEE, operator log-in, product and password management, and historical information on each machine—are often managed through data exchange with a database, which is simply an organized collection of data. Information is stored in a way that it can easily be accessed by categories at a later date. Pieces of data can relate to each other in various ways. Correlations of one type of information with another must be made in a meaningful way from the factory floor or from within management software.

Data is usually categorized in classifications that can support the process of relating the information pieces to each other and drawing conclusions from it later. An example might be relating machine faults to product selection to determine if a machine may need mechanical work on a station to accommodate a specific product.

Here is another example of how a database might be used in a business scenario. An engineer keeps a schedule of critical production and task dates for each job number. Purchasing uses a spreadsheet with component part numbers and delivery dates, also by job number. A project manager then uses a database to enter customer information and link to the engineer and purchasing data sets by using the job number as the primary key. Reports can then be generated using this information for a specific project, perhaps by determining whether a project schedule slipped because of late part shipments or other causes. By changing the primary key to look at reports by vendor, comparisons can be made to track on-time and correct part shipment percentages.

A database is technically only the data and its structures, not the management and relational aspects that control it. The engines that actually perform searches and access data are contained within the database management system (DBMS). The data and DBMS together are called a database system.

Commonly used database management systems include Microsoft Access and SQL Server, Oracle, IBM DB2, and several SQL-based variants. SQL is an abbreviation for Structured Query Language, a method of relating data categories in multiple ways. Object-oriented and object-relational databases often use a query language called Object Query Language (OQL), which uses many of the same rules, grammar, and keywords as SQL. A DBMS is required to manage data according to availability to multiple users simultaneously, accuracy, usability or user-friendliness, and resilience or recovery from errors. This can be a complex task, and systems often connect many different servers and data collections. Nearly every aspect of business uses databases in various ways and on different platforms that often do not communicate directly. This requires additional software to perform translation and control of data and its acquisition, usually using standards like SQL and ODBC together.

Databases can be classified in a number of ways—by their content, such as text, images, or file types, or by their application area, such as production, accounting, or maintenance. The term may also refer to the logical programming and data retrieval aspect or the data content in computer storage. Following are some of the different types of database in use.

A *relational database* stores data in rows and columns. Each row has a primary key that uniquely identifies each record in the table. It can either be a normal attribute that is guaranteed to be different for each record, such as a social security number, or it can be generated by the DBMS. Each column also must have a unique name. Access, MySQL, and SQL Server are examples of relational databases.

An *object-oriented database* (OODB) stores data in objects. Objects are items that contain data as well as the procedures that read or manipulate it. Members of the database are objects that might contain several related data characteristics, such as name, address, and age, as well as the instructions for how to print the record or a formula to calculate a member's paycheck. A relational database would contain only the data about the member. GemFire and Versant are two OODBs.

There are several programs that are known as object-relational databases, which have the ability to do both. Examples of these are DB2, Oracle, and Visual FoxPro.

A *multidimensional database* stores data in dimensions. While a relational database uses columns and rows (two dimensions), a multidimensional database allows for more than two dimensions, allowing users to access and analyze the view from any aspect. This multidimensional table is sometimes called a "hypercube." The number of dimensions used will depend on the requirements of the application. An example might be a database with product, manufacturer, vendor, time, and the model of machine it is used on. A user would be able to look up data by date, product number, vendor, and so on. Multidimensional databases are also faster at consolidating data than relational databases. A query that takes minutes to process in a relational database may take seconds in a multidimensional database. A well-known multidimensional database is Oracle Express.

Databases are also classified by their content. A multimedia database stores images, audio clips, and video clips. A voice mail system database is an example. A groupware database stores schedules, calendars, manuals, memos, and reports. Searching a schedule for available rooms or meeting times would use a groupware database. A CAD database stores data about engineering designs and drawings. It might include lists of components, relationships between parts or drawings, and drawing revision data.

When designing a database, it is a good idea to follow some basic guidelines. First, the purpose of the database must be determined. This will help the designer with what kind of information will be required. Next, the tables or files should be designed. Each table should contain data about one subject. For example, the product file should only contain data about the product. Third, the records and fields are designed for each table or file. Each record should have a primary key. Separate fields should be used for logically distinct items; a name, for instance, should have fields for title, first, last, and so on. Fields should not be created for information that can be derived from other fields (for example, the field for age can be derived from a birth date field). After the tables and files are completed, the relationships between them can be determined, completing the design.

For database users, macros can be used to remember sequences of operations, automating repetitive functions. These macros can be saved and reused any time a similar task needs to be performed, creating easily modified user tools.

6.7 Enterprise Software

Enterprise computing often involves the use of computers in LANs or wide area networks (WANs). Businesses produce and gather large volumes of information about customers, products, suppliers, and employees. This information often flows both inside and outside the enterprise, with users consuming information and computers tracking interactions.

Information systems can be organized into five basic categories:

1. *Office information system (OIS)*: This enables employees to perform tasks with computers rather than manually. This is sometimes referred to as office automation. OISs support administrative activities such as word processing, spreadsheets, databases, presentation graphics, e-mail, personal information management, and groupware. Appointment management and scheduling are also components of this system.

2. *Transaction-processing system (TPS)*: This captures and processes data from daily business activities. A transaction is an individual business activity, such as an order, deposit, payment, reservation, or employee clock-in. TPSs were some

of the earliest forms of computerized data processing. Early TPSs used batch processing; in this method the computer collects data over time and processes all transactions later as a group. Batch processing is used by many businesses to calculate paychecks or print invoices. Online transaction processing is done in real time, with data being exchanged immediately. Credit card processing is an example of this.

3. *Management information system (MIS)*: This generates information for managers and other users to utilize in making decisions, solving problems, supervising activities, and tracking progress. The computer's ability to compare data and generate reports with only the required information makes this a useful tool. MISs are usually integrated with TPSs. The system can take information on customer sales, account balances, and inventory and generate reports recapping daily, weekly, and monthly activities. It can flag issues like unpaid balances or spot trends and make forecasts. MISs create three basic types of report: detailed, summary, and exception. Detailed reports are simply lists of transactions, usually sorted by date. Summary reports make data easier to understand by consolidating it into tables, charts, and graphs with totals for some time period. Exception reports flag data that falls outside the norm, such as unpaid bills or part inspection failures.

4. *Decision support system (DSS)*: This helps users analyze data generated by TPSs and MISs. These systems often use data from both internal and external sources. Internal data might consist of sales figures, inventory, or financial data, while outside sources may include interest rates or economic forecasts. DSSs often contain statistical analysis tools, spreadsheets, graphics, and scenario-modeling capabilities. Executive information systems (EISs) are a special type of DSS that present information as charts and tables that show trends, ratios, and statistics.

5. *Expert system*: This is programmed with the knowledge of human experts and imitates human reason and decision making. Expert systems are composed of two main components: a knowledge base and inference rules. These rules are a set of logical decision algorithms that are applied to the knowledge base whenever a user describes a situation to the expert system. This is an example of the use of artificial intelligence (AI) and adaptive learning in computers.

Application software used in business makes use of all five of these different types of information systems by combining them into packages. A common term for this is *integrated information systems*.

Customer relationship management (CRM) software manages information and interactions with customers and prospects. It is used primarily across sales, marketing, and customer service departments within an enterprise. It tracks correspondence and sales and helps companies gain a competitive edge by using analytical tools. An important aspect of this software in today's market is social media and interaction with customers on mobile devices.

Enterprise resource planning (ERP) software provides a centralized and integrated platform for all the major business activities of an enterprise. This type of software is customized for different kinds of business and often takes years to fully implement. It integrates information across all departments, providing a complete view of the entire organization for management. Because information is shared rapidly, ERP helps manage global operations in real time. The reliance on one system rather than many from different vendors allows the IT (information technology) department to focus on one type of technology. ERP software packages can still be modular, with separate solutions for financials, human capital management, sales and service, procurement and logistics execution, product development and manufacturing, and corporate services. These can be purchased separately from the same vendor and integrated into the system as it grows. The largest and most well-known vendor for ERP and other enterprise application software is SAP.

Content management systems (CMSs) are a combination of databases, software, and procedures that organizes and allows access to a variety of different kinds of documents and files. It includes information about the data and files called metadata. This type of information might include revision numbers, a brief summary of the file, and the name of the author. The CMS also includes security controls that limit access to the file, adding content to the database, or modifying its content. Content is added through a GUI or a secure portal with a web page. The CMS provides the ability to categorize, index, process, and store content within the system.

Material requirements planning (MRP) software helps monitor and control processes related to production. It includes inventory management and forecasting tools to ensure that materials required for manufacturing are available when needed. Basic functions include inventory control, BOM processing, and scheduling tools. MRP is used to plan and integrate manufacturing, purchasing, and delivery activities. Data that is considered in MRP software includes shelf life of stored materials, BOMs, how many and when items are required for production, materials in stock and on order from vendors, and planned production goals. Output from the MRP software encompasses the recommended production schedule and the recommended purchasing schedule.

Manufacturing resource planning (MRP II) is an extension of material requirements planning. It includes elements of MRP but also contains tools to track production in real time and monitor product quality. Shop floor data collection may be done using manual data entry from operators or by integrating machinery control systems with the MRP and MRP II software. Likewise, quality control may be a combination of periodic product checks with manual data or real-time measurement and gauging integrated with production machinery.

Integration of plant floor systems and business enterprise software can be a complex task. Production control systems require quick response and deterministic processing. Data collection is often done within the machine control system itself and may be transferred periodically to the enterprise system or it may be performed on a continuous basis. Because of the bandwidth requirements of data exchange on the plant floor, it is usually best to separate the enterprise data collection systems from the control systems. This is often done by implementing several layers of communication networks with separate ports for each. Figure 6.11 illustrates how a plant network might be set up to isolate traffic and provide security for different communication layers.

Additional security measures employed in a plant data network include password log-in and administration that changes passwords periodically and ensures password strength. Communications are often encrypted and firewalls employed to prevent hacking. Virus protection is commonly provided for every computer on the network. Separation of networks as shown in the plant data network can also be effective for security. Plant procedures such as prohibiting USB or thumb drives can also have some limited effect.

Data historian software is essentially a database that stores historical information about processes or machinery. Data may be

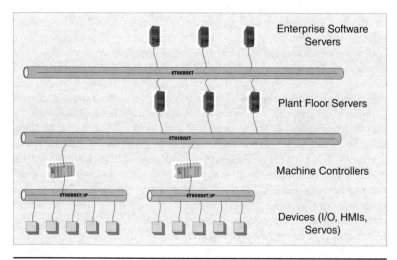

Figure 6.11 Plant data network.

updated on change of state of monitored tags, periodic updating of monitored process values, or by manual entry by an operator. Examples of data that might be stored include total products or total defects for a shift or day, current temperature of an oven, maximum flow rate of a pump over a period of time, or the reason for a line stoppage entered by a maintenance technician.

A *batch server* is used to help applications develop, schedule, manage, and monitor batch jobs as described in the previous TPS section (Sec. 6.7). Because batch jobs must often be load-balanced across several tasks, they are often placed on their own server in a network.

There may be many different servers on a network performing different tasks, all communicating with each other simultaneously. The load on the network can vary, network traffic tends to peak at certain times. Choices of hardware, network layout, and software implementation can all have a major effect on network speed and efficiency.

CHAPTER 7

Occupations and Trades

There are quite a few different kinds of employment opportunities in the automation field. Manufacturing facilities are becoming increasingly more automated to improve production, vendors are looking for knowledgeable people to sell their technical products, and OEMs employ skilled labor to design, maintain, and build their products. System integrators and machine builders hire engineers and tradespeople to assist companies in solving their automation problems.

7.1 Engineering

Engineering is a discipline that applies scientific knowledge in the fields of physics and chemistry, materials, mathematics, and logic to solve real-world problems. Engineers use the tools they acquire in the study of scientific and mathematical principals to invent, design, and create physical solutions to problems.

Creativity is a major factor in the application of science to the physical world. The design and development of structures, machines, and processes require a full understanding of the materials and physics of the components and devices used. Behavior of machinery and processes must be forecast under all operating conditions. Personnel and equipment safety, the economics of designing, building and operating equipment, and ethical practice of the engineering profession are important elements in the training of engineers.

Licensing requirements vary depending on the engineering discipline and its application. Formal designations include the Professional Engineer (PE) and Chartered Engineer license. The basic requirement for an engineer is the completion of a bachelor of science degree from an accredited university.

Engineering disciplines are divided into a number of subfields. There are basic math and science classes that all engineers must take, including calculus, physics, and chemistry. In addition, most engineering courses require a certain amount of cross-training from other disciplines as well as general college coursework.

7.1.1 Mechanical

Mechanical engineers design assemblies and systems to accomplish automated tasks. They are often hired to supervise a cell of machines in a manufacturing plant or upgrade a production line. Their job usually includes the use of design and CAD software, both two- and three-dimensional. Design processes will often involve timing charts to analyze machine movement and the relationships of components to each other.

Basic mechanical engineering coursework includes solid mechanics, instrumentation and measurement, strength of materials, hydraulics and pneumatics, combustion, and product design. Mechanical engineers also study fluid flow and thermodynamics. Cross-disciplinary training often includes some computer programming, electrical, industrial, and possibly chemical engineering classes. Mechanical engineers are often involved with the specification of sensors and the effects of temperature or chemicals on different materials or manufacturing processes. Specializations include robotics, transport and logistics, cryogenics, biomechanics, vibration, automotive engineering, and more.

Knowledge of components such as motors, bearings, linear actuators, gearing, and various other elements, as described in previous chapters, is critical. The ability to specify and size framing and piping components and select appropriate material and components is an important skill. Mechanical engineers and designers rely heavily on manufacturers' specification sheets and their own knowledge of a range of components. Vendor training can also be of great value.

7.1.2 Electrical and Controls

Electrical engineers are often involved in systems design and software for automation equipment. In manufacturing, they are usually involved in modifying code in the systems controller, specifying and adding sensors, modifying HMI screens, and adding motor or power circuits for line upgrades. In system integration or machine-building companies, controls engineers design electrical control panels and draw schematics, flowchart, and write control code. When designing machines, they work very closely with their mechanical counterparts for a fully integrated machine design. They are usually responsible for the start-up and debug of machines also.

The curriculum for electrical engineering includes the basic science and math coursework described previously. In addition, mechanical classes in statics and dynamics, thermodynamics, and materials science are usually required. Engineering economics and engineering ethics are also generally part of the program. After studying general electrical design and concepts, electrical engineers usually specialize in a subdiscipline, such as power, electronics, digital circuits and microelectronics, optics, controls, plasma engineering, communications, or computers.

For OEM equipment, knowledge of IC design can be important. Some OEM equipment uses proprietary circuit boards and "System on a Chip" components in the design of their equipment. Printed circuit board design and layout can be an important element in machine control design.

For process systems, electrical and controls engineers need to know about power distribution and P&ID diagrams. They also often end up in niche fields, becoming specialists in fields like robotics, vision, or integrated servo systems. A good foundation in IT skills is often important as plants become more integrated between the factory floor and production planning and management. Software and computer programming is a required skill for all electrical engineering disciplines.

Because controls engineers often take the lead in the start-up of equipment, they must also have a good mechanical background. Pneumatic design and specification are often performed by electrical or controls engineers. Motor sizing involves the evaluation of mechanical systems and machine dynamics.

7.1.3 Industrial and Manufacturing Engineering

Industrial engineers determine the most effective way to use people, equipment, and materials to produce a product. They are often involved in planning and efficiency studies of complex systems and production procedures. Ergonomics and safety are also areas that industrial engineers often have responsibility for. Although not as often involved in the functional design of machinery, industrial engineers are usually the ones who decide where machines will be located in relation to one another for the most efficient process flow.

Material movement, flow, and storage are also issues addressed by industrial engineers. Processes are often a combination of manual and automated procedures that must be analyzed and optimized. Industrial engineers are often trained in Six Sigma process improvement and other business-oriented disciplines. Operations management is one of the underlying concepts behind industrial engineering. As such, a good knowledge of commercial enterprise computer software and platforms is important. The use of quantitative methods and statistics in the analysis of industrial operations is an important tool for industrial engineers.

The industrial engineering curriculum includes the essential math and science classes required for all engineering disciplines in addition to specialized courses in management, systems theory, ergonomics, safety, statistics, and economics.

Manufacturing engineers direct and coordinate processes for production. They are involved with the initial concept of how a product will be produced through the full implementation of a production line. The manufacturing engineering curriculum is very similar to that of a mechanical engineer. Manufacturing engineers are often formally trained in another discipline, such as industrial or mechanical engineering, but are designated manufacturing engineers because of the position they occupy in the industry.

The Society of Manufacturing Engineers (SME) provides certifications for manufacturing engineers in the United States. Candidates for the Certified Manufacturing Technologist (CMfgT) certificate must have four years of combined education and manufacturing-related work experience. A three-hour, 130-question exam covering mathematics, manufacturing processes, automation, and manufacturing management is administered to qualified candidates for the certification. The Certified in Manufacturing Engineering (CMfgE) qualification requires eight years of combined education and manufacturing-related work experience. A passing grade in a three-hour, 150-question exam, which covers more in-depth topics of the CMfgT, is required for this certification.

SME also provides additional certifications for engineering management, Six Sigma, and lean manufacturing. Engineers from many different disciplines often obtain these certifications in addition to their formal training.

7.1.4 Chemical and Chemical Process Engineers

Chemical engineering involves the design and operation of plants and machinery in the chemical- and bulk-processing areas, as well as converting raw materials and chemicals into other forms. These disciplines are often subdivided into chemical process engineering and chemical product engineering. The processing of materials in solid, bulk, liquid, and gaseous forms is the focus of process engineering, while the individual unit reactions of substances and elements with each other for commercial use is the subject of product engineering.

Chemical reaction engineering concerns the management of processes and conditions to ensure a safe and predictable chemical reaction. Models are used to simulate processes and predict reactor performance. Process design involves activities such as drying, crystallization, evaporation, and other reactant preparation steps. Conversion processes are also designed for nitration, oxidization, and other material effects. These involve biochemical, thermochemical, and other processes. Transport of materials involves the effects of heat transfer, mass transfer, and fluid dynamics as substances or compounds are moved from one place to another.

In addition to standard engineering course requirements, chemical engineers take various science and engineering courses, including physical chemistry, organic chemistry, biology, biochemistry, reactor design, reactor kinetics, fluid flow and thermodynamics, statistics, instrumentation, and environmental engineering classes. Chemical engineers are involved in designing and optimizing processes for commercial product production. As such, they need a good background in mechanical and electrical disciplines. Process engineering involves the application of heating and cooling, pressure and vacuum, bulk movement, design of reactor vessels, and piping. P&ID diagrams are used to describe the processes in terms of components and fluid flow/airflow, which chemical engineers are usually familiar with.

In industrial automation, petrochemical-related processing is a major employer of chemical engineers—both in the extraction and refining fields and in plastics. Biological sciences, waste management companies, and pharmaceutical manufacturers also commonly hire chemical engineers. Most process-oriented companies, such as nonwoven web converting, paper manufacturing, chemical compounding, and consumer product manufacturing firms, also use chemical engineers for process and product design.

7.1.5 Other Engineering Disciplines and Job Titles

Within most companies that use or implement automation, there are usually a variety of job titles associated with engineering. Plant managers, for instance, are often from the engineering field, since problem solving is an important skill for both job functions.

Quality engineers may come from another engineering discipline also, although industrial and systems engineering programs teach many of the elements of quality programs. Six Sigma and lean manufacturing techniques are used by quality engineers to make processes more efficient and reduce defects. Total quality management (TQM) is a technique used to continuously improve a product or service. The production part approval process (PPAP) is another element of quality engineering.

Systems engineering is an interdisciplinary field of engineering that focuses on how complex engineering projects should be designed and managed over the life cycle of a project. Issues such as logistics, the coordination of different teams, and automatic control of machinery become more difficult when dealing with large, complex projects. Systems engineering deals with work processes and tools to handle such projects, and it overlaps with both technical and human-centered disciplines, such as control engineering, industrial engineering, organizational studies, and project management. Some universities offer advanced degrees in systems engineering. Systems engineers are generally employed by large companies or government-related (DoD, DoE) manufacturers.

Applications

Applications engineers do a lot of the pre-engineering on projects. The task of an applications engineer is to come up with a way to perform a given task in the quickest or most cost-effective way. Vendors and manufacturers often employ applications engineers to support the sales staff and provide value to the buyers of their products. They also often perform technical training in this role.

Machine builders and integrators use applications engineers to put together quotes for machines and systems. They will often use in-house developed tools to estimate costs for parts and labor and 3-D software to develop layouts and concepts for machines or process lines that will be embedded in quote documents.

Applications engineers often spend a lot of time visiting plant sites and seeing how things work. They usually have many years of experience and often come from a design background.

Sales

Sales engineers are usually employed by vendors and manufacturers of automation equipment. They often also act as applications engineers, helping customers determine the best application of their technical product. Sales engineers often attend factory training on their products and can be of immense help to design engineers in coming up with the best way to solve a particular problem.

Sales engineers often carry or have access to samples of their product, which can be tested or examined by the customer. They also may have software demos that can be used to illustrate the application of the product. It goes without saying that people skills and good written and verbal communication are important parts of a sales engineer's toolbox.

Training programs in sales techniques and sales management are available from many third-party companies that specialize in the sales training field. There are also many excellent books available. Public speaking and communications skills training are also available through associations and companies.

Project Engineers

Project engineers usually perform a lead role in the actual implementation of an automation project. Not only are they responsible for the overall design; they also often have to interface with the customer regularly. Mechanical and electrical project engineers typically team up to implement a quoted project. As such, they are often well versed in each other's disciplines.

Project engineers usually have experience as a design engineer prior to taking project responsibility. They are usually proficient in CAD or design software and knowledgeable about many products.

Design Engineers

Design engineers work with CAD software within their discipline to create mechanical or electrical schematics for fabrication. They need to be adept at their design software platform and able to examine designs with a critical eye for possible mistakes. Mechanical design engineers usually work with 3-D modeling software to create solid models that can be used to simulate machine motions. Drawings are often then converted to 2-D printouts of components for fabrication or converted to a CNC compatible file, such as G-code or STEP-NC.

Electrical design engineers take design elements, such as mechanical layouts, I/O lists, and machine specifications, and convert them into electrical schematics. They are also involved in the specification and selection of electrical components. Along with project engineers, design engineers are responsible for ensuring that specifications for an automation project are followed.

Project Managers
Project managers are responsible for adherence to the schedule and budget of a contract. While they often come from an engineering background, they usually have experience in a business field also. Financial spreadsheets and scheduling software are important tools of the trade. Interfacing with the customer and generating issues lists, working with purchasing for expediting, and keeping management apprised of project status are also important project management tasks. Project managers often perform supervisory tasks for teams of project engineers and act as a liaison between the customer and the engineering team.

Project management certification and training often includes many of the lean manufacturing and Six Sigma techniques discussed later in this book. The Project Management Institute (PMI) administers the certification of project managers through training and testing. Much of a project manager's core training information is contained in the *Project Management Body of Knowledge* (PMBOK), authored by the PMI. Certifications for project managers include Project Management Professional (PMP), Program Management Professional (PgMP), PMI Agile Certified Practitioner (PMI-ACP), PMI Risk Management Professional (PMI-RMP), and PMI Scheduling Professional (PMI-SP).

7.2 Trades

The physical implementation of automation equipment and machinery is usually done by skilled tradespeople with extensive experience in their field. Often, people who work on equipment have to make "design-on-the-fly" decisions as a machine is being built or modified. This requires knowledge and experience in a variety of fields, including fabrication techniques, materials, and sensor applications.

7.2.1 Mechanical

Manufacturing and Machining
Machinists and operators of programmable CNC equipment produce components for automated machinery. They are usually trained on a

variety of equipment, ranging from mills, lathes, and grinders to water cutting and sheet metal forming equipment. They work off detailed prints or "details" provided in CAD format from engineering. In the case of highly automated CNC machinery, they have to be able to program the machining center to produce the desired part.

Other titles and job descriptions that are related to machinists include tool and die makers, mold makers, pattern makers, and others. A person who produces mechanical parts is sometimes referred to as a "turner," while a person who assembles them together can be referred to as a "fitter."

Setup of machining equipment is critical to the precision and accuracy of the finished part. Jigging and fixturing contributes to a machine staying within the tolerances that a part is designed to fit. If parts are not set up and fixtured properly, expensive material can be destroyed. Precise measurement of parts using a variety of tools and instruments is an important factor in the accurate fabrication of parts. The use of tools, such as micrometers, calipers, and coordinate measurement machines (CMMs), is an integral part of a machinist's training.

Formal training programs for machinists are usually offered by vocational schools and community colleges. Two-year degree programs in machine technology focus on theory and technical skills. Classes include lathe and milling operations, CNC machining, precision measurement, blueprint reading, math, and quality control. Trainees often apprentice in machine shops under the supervision of skilled machinists. Certifications for machinists include degrees from accredited programs and testing from societies, such as the Fabricators and Manufacturers Association (www.fmanet.org).

Knowledge of the properties of the materials they are working with is essential to a good machinist. Metals, such as steel and aluminum, as well as materials such as UHMW, Delrin, or Teflon, all have their own associated techniques required to form them as well as associated tool speeds. Knowledge of treatments, such as anodizing, heat-treating, and forming, is commonly used to shape or change the properties of materials.

Assembly

When building machinery, assemblers put together the various components of a machine according to a set of prints. Aluminum extrusion is often used for guarding—or even for the whole machine—and knowledge of different types of connections, such as brackets, end fasteners, and anchor fasteners, is important. A variety of fastening techniques, such as using bolts, screws, dowel pins, or welding, is employed to mount parts to the frame of the machine. Mechanical actuators and linear or rotary motion components also each have their own associated assembly techniques.

Pneumatic and hydraulic plumbing and routing, conveyor assembly, and even a machining background can all be helpful skills in this field. Assemblers are often also skilled in machine wiring.

When machinery has already been built, millwrights or other maintenance personnel generally perform other tasks associated with working on automation equipment.

Welding

Machine frames and piping systems are often welded for stability. Welding techniques are discussed in depth in Chap. 5. Stick welding or SMAW is a technique that cannot be easily automated and is only applicable to manual welding. MIG, or wire welding, is also often used in manual applications.

Other techniques closely related to welding include plasma cutting, brazing, heat control and metallurgy, test methods, and safety. nondestructive evaluation (NDE), also known as nondestructive testing, is often used on welds to check for lack of fusion of the weld to the base metal, cracks or porosity within the weld, and variations in weld density. Techniques for testing include X-rays, ultrasonic testing, liquid penetrant testing, and eddy current testing. Welders should be familiar with these techniques.

Welders often attend training schools and require certification for their trade. The American Welding Society (AWS) has certification programs for testing procedures used on structural steel, petroleum pipelines, sheet metal, and chemical refinery industries. AWS can also test to company supplied or noncode welding specifications. Certification programs include the Certified Welder Program (CW), Certified Welding Engineer Program (CWENG), and Certified Welding Inspector Program (CWI). Certification is also available for robotic arc welding. Welding is subject to inspections and testing on a per-piece basis and costs more if done improperly. Like machinists, welders also require a solid background in the properties and behavior of metals.

Millwrights

When moving and installing large machinery, millwrights and riggers often perform most of the work. Operating equipment like forklifts and cranes is an important part of this. Their job requires a thorough knowledge of the load-bearing capabilities of the equipment they use, as well as an understanding of blueprints and technical instructions. Another common name for a millwright is a *rigger*.

Millwrights must be able to read blueprints and schematic drawings to determine work procedures and to construct foundations for and to assemble, dismantle, and overhaul machinery and equipment. They must be able to use hand and power tools and direct workers engaged in such endeavors. The use of lathes, milling machines, and grinders may be required to make customized parts or repairs. In the course of work, millwrights are required to move, assemble, and install machinery and equipment such as shafting, precision bearings, gearboxes, motors, mechanical clutches, conveyors, and tram rails, using hoists, pulleys, dollies, rollers, and trucks. Additionally, a millwright may also perform all duties of general laborer, pipefitter, carpenter, and electrician. A millwright may also perform some of the duties of a welder, such as arc welding, MIG welding, and oxyacetylene cutting.

Millwrights are also involved in routine tasks, such as machinery lubrication, bearing replacement, seal replacement, cleaning of parts during an overhaul, and preventative maintenance.

7.2.2 Electrical

Panel Building

Panel building involves the mounting of electrical components to the backplane of a metal enclosure and the wiring of these components. Panel layout drawings and schematic diagrams are used to lay out the components in a prescribed manner. Adherence to the National Electrical Code and customer specifications is a critical part of the process. Electrical panels take many different forms and contain a variety of different types of components. Elements—from PLCs and other controllers to motor starters and servo drives—may be mounted in the panel. Voltages from 5 or 24VDC to 480VAC or higher may be present in the same enclosure, and great care must be taken to keep these separated.

Mounting techniques include riveting, drilling, tapping, and cutting or punching holes in the metal enclosure to mount rectangular components. Wiring involves labeling of wiring conductors, use of ferrules to prevent the splaying or birdcaging of stripped wires, and termination of wires and cables to terminal blocks or component terminals. Spade and ring terminals are often crimped to the ends of wires for termination.

Soldering is an important skill for a panel builder. Wires may be attached to circuit boards or plug connectors this way, and a good connection is critical.

Routing of wires and cables through wireway— typically plastic with "fingers" or tabs to allow routing through the sides—is an important wire management element. Ensuring that components are mounted straight, labels are legible and in the same direction, and wires are formed into neat bends is another important aesthetic part of panel building. A panel-building operation in progress is shown in Fig. 7.1

Tools used by a panel builder include wire strippers, crimpers, screwdrivers, and a host of other hand tools. There are also various special-purpose tools, including hole punches, DIN rail shears or

Occupations and Trades

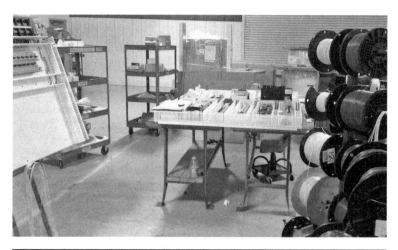

FIGURE 7.1 Panel building.

cutters, and label printers that are specific to wiring and control panels.

Panel builders typically learn their craft through experience or training programs at some large manufacturers. Unlike with machinists, welders, and electricians, there are no formal training schools for panel building.

Electricians

The electrician trade involves wiring of buildings, machines, and process equipment. This is differentiated from linemen, who work on electrical utility distribution systems at higher voltages. Electricians often concentrate in three different general categories: residential, commercial, and industrial wiring. In the industrial automation and manufacturing fields, electricians fall into the industrial grouping. Industrial electricians are responsible for facility wiring and maintenance and often work on high-voltage distribution bus bars mounted on the ceiling. Knowledge of factory power distribution systems and switchgear is part of an industrial electrician's tool kit.

Electricians typically perform the external wiring of machinery. They route wiring from the control enclosure to the various motors, sensors, and other electrical devices on a machine or line. This wire or cable may be run in conduit, emt, cable tray, metal wireway, or even directly strapped to the frame of the machine, depending on the requirement.

Training for electricians usually involves an apprenticeship under the general supervision of a master electrician and the direct supervision of a journeyman electrician. Electricians often attend two years of vocational school or community college for training in electrical theory and the National Electric Code. Electricians are usually licensed, but for wiring of machines, this is usually not a requirement. Licensing is generally performed in the United States at the state level, while enforcement is generally undertaken on a local level.

Electricians may often build control panels, but panel building and machine wiring are entirely different skills. In an automation environment, electricians must also be familiar with pneumatics and hydraulics, as they may be the ones who route air hoses and hydraulic piping on a machine or production line.

Knowledge of the National Electrical Code, wiring techniques, and devices and a familiarity with various hand tools are necessary skills for electricians. Wire strippers, cable cutters, and multimeters or volt-ohm meters (VOMs) are standard tools for electricians. Knowledge of electricity and power distribution, as well as various conduit bending and connecting techniques, is important. In addition to safety issues associated with working at heights and with power tools, electricians must always be aware of the hazards of working with electricity.

Instrumentation Technicians

In process control facilities, instrumentation is a key element in the monitoring and control of production. Maintenance and calibration of devices and troubleshooting of systems is the job of specialized technicians well versed in electronic, pneumatic, and hydraulic systems. Centralized monitoring of control loops is often accomplished with DCSs. Signals are wired to and from this central point from widely separated locations. This wiring network often involves intermediate control and junction points as well as communication-network-based I/O.

Local display of pressure and flow is often accomplished with mechanical gauges. Some gauges must be placed in-line with the process flow, while others are plumbed parallel and may have pneumatic tubing or piping interfaces. Mechanical interfacing is almost always an element of the instrumentation process.

Technicians must be able to interface with valves and instrumentation through SCADA, HMI, or other monitoring software to determine the causes of process problems. The ability to read and modify P&ID diagrams and electrical schematics are important skills. Calibration of instrumentation must be accomplished on a periodic schedule and records carefully maintained. Mechanical aptitude is also important, as pipefitting and welding can often be required. Instrumentation technicians and other process control personnel often undergo extensive safety training and usually require certification. Training programs are generally available in community colleges, vocational schools, and larger process-focused corporations.

CHAPTER 8

Industrial and Factory Business Systems

There are many different types of businesses involved in the field of automation in one form or another. Manufacturers and OEMs use automation components in the production of goods, manufacturers' representatives and distributors sell automation components and devices to OEMs and end users, and machine builders and systems integrators use and specify components to produce machinery and systems.

Functions within these companies can vary widely depending on their size and organization, but they generally fit within the same general structure.

8.1 Automation-Related Businesses

There are many different types of businesses that produce or make use of automated equipment. Government agencies, such as the Department of Defense, Department of Energy, and U.S. Department of Agriculture, all use automation equipment. The following is a list of broad categories of automation-related businesses.

8.1.1 Manufacturers

The largest users of automated production equipment are manufacturers. Manufacturers may produce individual products or process materials, as described in the converting section of this book (Sec. 5.4). Larger manufacturers may also build their own automation equipment since they have developed the expertise within their fields over years of experience.

Of course, some manufacturers also produce automation components and devices themselves. A list of some of these manufacturers is included in the back of this book (Appendix E).

8.1.2 OEMs

OEMs are manufacturers that produce machinery or equipment of their own. These may include standard products, such as appliances or automobiles, which include automation components. Standard or niche automation-related products made by OEMs include vibratory bowls, torque drivers, compressors, cam indexers, web-processing equipment, testers, ovens, and many other system components.

Many OEMs build standard products that are controlled by PLC or DCS. Examples of these are building HVAC and environmental controls, food-processing machinery, and packaging machines. Others may control their product with embedded processors or "systems on a chip."

8.1.3 Manufacturers' Representatives

Manufacturers' representatives serve as a regional sales and technical resource for automation component and device manufacturers. They often handle multiple products and call on both end users and distributors. They provide product demonstrations, training, and seminars. Typically a manufacturers' representative's income is a percentage of all products sold within his or her territory. For some specialized products, such as instrumentation, the representative may sell directly to the end user.

Manufacturers' representatives are often used by smaller companies that cannot afford to have a full-time salesperson within a territory. In this case, the representative is expected to prospect and follow up leads within his or her area. Other representatives may work directly for the manufacturer and cover a large geographical area. In this case, they often evaluate applications and perform training.

8.1.4 Distributors

Distributors sell products directly to end users. They usually fall into general categories, such as fluid power, electrical, or industrial supplies. Often distributors have branch offices over a region or nationwide. A distributor with a physical location who keeps product on the shelf is known as a stocking distributor.

Most distributors do not have exclusive rights to a product within a geographic area but compete with other distributors. Exceptions to this are Allen-Bradley and Siemens, who generally only allow one distributor in each territory. In exchange for this exclusivity, they attempt to control pricing and require the distributor to employ a technical support staff.

8.1.5 Machine Builders

Machine builders are different from OEMs in that they build machinery customized for a specific purpose. Although they sometimes specialize within a field, such as material handling or inspection and gauging, they will often take on any project they feel comfortable executing.

Machine-building companies can vary in size from four or five people building small machines to large firms with hundreds, or even thousands, of employees spread over multiple locations. Larger machine builders house various departments, as described in Sec. 8.2.

Machine shops often branch out into machine building. They may form a relationship with a systems integrator or controls company to design and build machines.

8.1.6 Systems Integrators

Systems integrators take separate systems and combine them into a functioning entity that operates as a whole. They are usually oriented around controls and IT but often have mechanical capabilities or partnerships with machine builders.

A company that does programming and/or panel building is often referred to as a "controls house." These companies are typically fairly small and do systems integration also. An example of systems integration on a small scale would be integrating test equipment, such as leak testers or gauging systems to a machine control system. This usually involves displaying results from the test and overall machine status on a touch screen.

Larger integrators often focus on specific industries, such as wastewater, environmental control, or chemical processing. Others may take on any type of job above a certain dollar figure. Closely related to systems integration companies are engineering firms. These companies usually perform some systems integration tasks and subcontract others. They usually employ professional engineers that can stamp or certify documents for large projects.

8.1.7 Consultants

Consultants are often used to provide expertise in a specific area of a project. They may have expertise in business enterprise systems, lean/Six Sigma, or a technical specialty, such as vision, data acquisition, or software. They may charge by the hour or week, or they can be kept on a retainer.

8.2 Departments and Functions

Business organizations usually have three basic functional areas: operations, finance, and marketing. Automation and manufacturing activities fall within the area of operations, which is responsible for producing goods and services for consumers.

Most of the companies involved in the industrial and automation fields are incorporated businesses with a variety of departments. Companies may be public (owned by stockholders) or privately held.

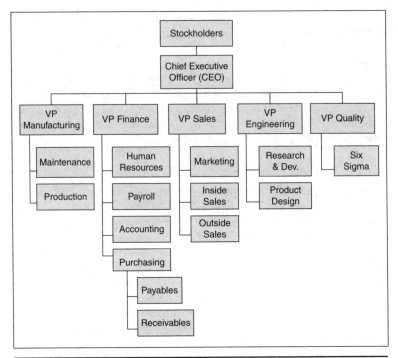

FIGURE 8.1 Organizational framework.

The functions of the departments may differ depending on the size and type of business. Larger companies and corporations tend to have more layers of management. A business with fewer layers of management is said to have a "flatter" organizational structure. Some large conglomerates separate business activities into completely independent organizations, each with its own structure. Manufacturing plants and product divisions are often organized this way, with little day-to-day interaction with a corporate headquarters on operational issues.

8.2.1 Management

There may be various layers of management, depending on the size of the organization. Usually there will be a single executive-level president or CEO (chief executive officer) with final decision-making authority. If the organization is large enough, there will be various layers of management below this in charge of each department or function. Titles for these may range from vice president to director or manager. Managers at the VP level are likely to be specialists in several areas. Managers at the CEO, president, and vice president levels are often referred to as upper management.

Below the upper-management level is middle management. This includes plant managers and department heads. For larger organizations, this second tier of management may operate in geographically different locations. Large manufacturers often have plants in different states or countries with management reporting to a headquarters. These functions may be fairly autonomous, requiring little oversight. Plant managers often have final decision-making authority on personnel and capital projects up to a certain cost.

Department heads have responsibility for such areas as production, maintenance, quality, personnel, accounting, and marketing. They may report to both the plant management of their facility and to the vice president of their functional area.

Lower management is responsible for direct supervision of production and maintenance workers. Job titles include line supervisors, foremen, or lead. Managers in this position have extensive experience in their craft or on the production equipment they supervise. They are often promoted from within the ranks of experienced machine operators.

8.2.2 Sales and Marketing

All businesses need ongoing efforts to present their products or services to customers. The vice president of sales generally leads this department for larger companies and may have several department heads under his or her supervision.

Marketing materials may include television and radio advertisements, printed brochures or flyers, and social network marketing campaigns. An Internet presence is an important part of marketing. Web sites provide information about a company's products and services and must be maintained and updated after they are established.

Sales efforts may involve outside salespeople who call directly on customers. Outside sales personnel demonstrate products, provide information, and maintain a relationship between companies. They also seek out new customers and follow up on sales leads. For highly technical products or systems, they often must have an engineering background or degree. Inside salespeople take orders from customers online or by phone. In some cases, they may assist outside sales with prospecting or customer research. Applications and technical support staff members help solve customer problems and answer product-related questions. Training seminars and classes are used to familiarize customers with company products and improve the perception of the product.

8.2.3 Engineering and Design

For businesses involved in manufacturing, the engineering department is responsible for product design, making changes to the product, ordering and installing new production equipment, and generally taking care of technical changes related to machinery. The department usually falls under a vice president of engineering or engineering manager, depending on the size of the company. Product or design engineers are responsible for the creation of drawings for the products themselves as well as tooling used in the manufacturing process.

Process-control-based companies also often have large engineering departments that design and install their own equipment. Since much of this equipment is customized to the process, it is often more cost-effective to modify a process internally. Many of the products used in process control are commercially available or easily outsourced to vendors, such as reactors and pressure vessels. Design engineers in these departments often design components and piping in a modular manner and then contract fabrication and installation services.

Within the organizational framework, there is also often a research and development (R&D) department or team that may operate under a separate VP. The R&D process involves design and testing of new products or modifications to existing ones. They may also design new machinery and processes to produce things in a different way. Whereas some research and development endeavors are focused on the development of products, others may attempt to discover new knowledge of scientific and technical topics to uncover opportunities for products and services that do not yet exist.

The PPAP is a standard used mainly in the automotive industry to ensure that component parts are made in a consistent manner and meet the user's needs. As part of this process, the design, manufacture, and testing techniques of a manufacturer are examined and documented. There are 18 elements within this process, extending from the initial design through qualification of production machinery and sample production parts. Two of these steps involve a standardized tool called failure mode and effects analysis (FMEA).

The design FMEA (DFMEA) evaluates the potential causes of failure within a part on a per-component basis. It identifies causes, establishes severity levels, and relates them to other components. As potential problems are identified, this gives designers an opportunity to correct problems. A process FMEA (PFMEA) considers the same elements in the manufacturing process. The PFMEA follows the steps defined in the process flow and determines the effects of failures in machinery and worst-case consequences for products, equipment, and personnel. This evaluation is of special concern to machine builders as they must comply with results of the analysis.

8.2.4 Maintenance

The maintenance department takes care of the equipment and facilities. Also known as maintenance, repair, and overhaul (MRO), personnel in this department are responsible for mechanical, plumbing, or electrical repair of the building and production equipment within it. The maintenance department is usually a subdepartment under manufacturing. Facilities and equipment maintenance may be organized into different departments.

Routine maintenance activities are separated into scheduled maintenance and preventive maintenance. Scheduled maintenance includes activities that keep equipment in working order by replacing worn devices and assemblies before they fail. Preventive maintenance involves activities such as lubrication, cleaning, and other actions that prevent age-related equipment breakdowns. Shutdowns are often held annually or semiannually to accomplish major repairs or upgrades; installations are also common during these periods. Machine breakdowns (unscheduled maintenance or repair) and upgrades must also be handled as required.

MRO software can be useful in managing scheduled maintenance activities. This tool can help increase system availability and uptime, track machine BOMs, and schedule routine maintenance tasks. It also creates a record of activities that can be useful for production. MRO software tracks component data, such as "as designed," "as built," "as maintained," and "as used." It can be used to manage repair inventory, store serial numbers, keep track of warranty and guarantee documents, and track machinery service history. MRO software is often integrated with other enterprise business software.

Total productive maintenance (TPM) is a program for improving machine availability. Equipment and tools are put on proactive maintenance schedules that involve operators in the maintenance of machinery. An aggressive maintenance schedule will include "deterioration prevention" techniques that identify issues as soon as possible and attempts to prevent problems from occurring. With added operator involvement, maintenance technicians are liberated from mundane maintenance tasks and enabled to focus on urgent repairs and proactive maintenance activities. TPM techniques rely on the 5S system used in lean manufacturing.

8.2.5 Manufacturing and Production

In a manufacturing company, most of the personnel are in the production department, which is usually under the control of a vice president of manufacturing, who generally reports to the CEO. When facilities are located in multiple geographic locations, plant managers usually report to the VP of manufacturing. Other subdepartments in the manufacturing sector include production control and scheduling, shipping and receiving, tool design, maintenance, and various industrial engineering functions.

Manufacturing engineers are often placed in charge of the installation, operation, and improvement of production lines. As such, they spend a lot of time on the factory floor evaluating machinery and production efficiency. They are often also involved in the supervision of production activities. Production supervisors oversee the stocking of raw materials and components required for the manufacture of products. They also interface with maintenance in ensuring that equipment is kept in safe working order.

Production employees often operate production machinery that can be quite sophisticated. Consequently, they are often trained in the care and maintenance of machinery and systems. Machine operators may be responsible for the upkeep of their assigned machine. Since they spend most of their working hours with the machine, they often get to know its quirks better than maintenance or even the company that initially built the machine.

8.2.6 Finance and Human Resources

The finance department includes purchasing, payables, receivables, payroll, and accounting. These are often separated into subdepartments, which report to a VP of finance or a comptroller who oversees accounting activities and auditing of company finances.

Purchasing directors or managers are responsible for the acquisition of goods and services. One of the jobs of the purchasing department is to obtain goods at the most advantageous terms to the company. This requires negotiation skills on the part of the buyer. Some products have a standard pricing arrangement that is nonnegotiable, but machinery and contracts are often sent out for bid. Evaluation of proposals and offered services are part of a buyer's skill set. Some purchasing agents set up "blanket" or "master" agreements for longer-term consumables in order to reduce the administrative costs of repetitive orders, also requiring negotiation.

Organizations often use a checks and balances system to ensure that the system remains ethical, often having different acquisition activities report to different senior managers. Departments involved in the buying process can include purchasing or buying, receiving, engineering, accounts payable, or plant management. As a two-way check if purchasing orders are issued, accounts payable will often process the invoice independently, checking with the requisitioner to ensure they received their order.

Accounts receivable is responsible for billing a company's customers for products sold and delivered. Accounts payable pays vendors for goods and services purchased. These functions often fall within the purchasing department but, as mentioned previously, may report to different supervisors. Payroll is a function that falls within the realm of finance but interfaces with human resources.

The purchasing department issues purchase orders to vendors for raw materials, components, machinery, and other items. Technical items are usually selected by engineering or maintenance staff; the vendor for these items, however, may be selected by purchasing. This can cause problems if the vendor has provided a service in the expectation of receiving an order.

The human resources department is responsible for the well-being of the company's employees. They handle employee benefits, such as health insurance and 401k and benefit plans. Attracting new employees, selecting and assessing them, and sometimes terminating employment are all duties of the human resource management (HRM or HR) department. The department is usually under the supervision of a VP-level personnel director.

Training usually also falls under the direction of HR. Employee training and education through the company may be handled internally or outsourced, depending on the subject matter. Companies also sometimes outsource human resource functions or use consultants. The human resources department is also usually responsible for labor relations in unionized plants. Collective bargaining agreements are negotiated between the union and the HR department.

The human resources department often administers employee benefit plans or payroll. These programs may be outsourced or run by specialists within the organization. HR is also often involved in the mergers and acquisitions process.

8.2.7 Quality

The quality department is responsible for the output of the company product in terms of individual product conformance to standards, quantity of scrap material, and efficiency of the process. Quality can be defined in various ways, depending on the product, but one way to think of it is the degree to which performance of a product or service meets or exceeds customer expectations. Quality control and quality assurance are usually placed under the direction of a vice president of quality. This department often has decision-making responsibility that supersedes that of the manufacturing and production departments.

A quality management system (QMS) is an organizational structure containing procedures, processes, and resources to implement quality management. Historically, quality control involved finding the defects in manufacturing before they left the factory. The methodology has changed extensively in the past 50 years or so because of implementation of standards and competition from foreign and domestic manufacturers. Personnel within the quality department are expected to be well trained in modern quality techniques and often specialize in specific areas.

Standards

There are many quality management techniques that have been developed and formalized for manufacturing. Standards such as ISO 9000 require that companies are audited for compliance periodically. ISO 9001 certification requires that a company is committed to the methods and model defined within the standard. This means that procedures for manufacturing and quality control have to be developed, documented, and followed, which requires time, money, and paperwork. This, in turn, can lead to fairly large quality departments within corporations that are required by customers to be certified. ISO 9000 standards include system requirements, management requirements, resource requirements, realization of requirements, and remedial requirements.

Another standard that companies often must comply with is ISO 14000. This standard concerns what an organization does to minimize harmful effects to the environment caused by its operations. The standard focuses on three major areas: management systems, operations, and environmental systems. Certification for this standard also falls within the domain of the quality department.

Total Quality Management

TQM is a philosophy that involves everyone in the organization in an effort to improve quality. There are many programs and techniques that have evolved from this philosophy that directly impact business operations and methods. There are a number of TQM practices that are identified within a wide variety of resources, including training courses and books. These practices include continuous improvement, involving everyone in the organization, and meeting or exceeding customer expectations. Cross-functional product design, process management, supplier quality management, strategic planning, and cross-functional training are also elements of the TQM philosophy.

The TQM concept was originally developed in the United States in the mid-1900s by several management consultants but was not widely accepted. The philosophy was adopted by Japanese automakers in the 1980s, and a number of new concepts were added. Among many other Japanese terms, *kaizen*—a Japanese term that means continuous improvement—made its way into the terminology of quality management as the philosophy evolved.

Six Sigma

Six Sigma is a business management strategy originally developed by Motorola in 1986. As of 2013, it is widely used in many industrial sectors. Statistically, Six Sigma means that there will be no more than 3.4 defects per million opportunities in a manufacturing process. From a quality perspective, the term refers to improving the quality of process outputs. Six Sigma achieves this goal by identifying and removing the causes of defects and minimizing variability in manufacturing and business processes.

Six Sigma uses a project methodology called DMAIC, an acronym for Define-Measure-Analyze-Improve-Control—a method used for improving existing business processes. In each step of DMAIC are several tools that can be used to achieve the next step of the methodology. Another tool used for creating new product or process designs is Define-Measure-Analyze-Design-Verify (DMADV)—also known as Design for Six Sigma (DFSS). Statistical tools used in the Six Sigma methodology include histograms, Pareto charts, regression analyses, scatter diagrams, run charts, and analyses of variance.

Six Sigma creates a special infrastructure of people within the organization composed of Master Black Belts, Black Belts, and Green Belts who are experts in these methods. Executive management is a critical component of an effective Six Sigma project, and "buy-in" at this level is a necessity. Each Six Sigma project carried out within an organization follows a defined sequence of steps and has quantified financial targets (cost reduction and/or profit increase).

8.2.8 Information Technology

The IT department takes care of the company computer systems related to both hardware and software. It spans a wide variety of areas that include computer software, information systems, computer hardware, databases, security, and training. Depending on the size and structure of a company, the manager of the IT department may be at a VP level (VP of information technology or chief information officer [CIO]) or at a network administrator level.

IT provides businesses with several types of service that help execute business strategy. The department maintains business enterprise, office, and CAD software, including associated databases and security. It maintains and improves existing network and hardware systems, and it is often necessary for IT personnel to write programs and software utilities to integrate existing software and databases.

Security is an important part of a company computer network. The IT department is responsible for developing and enforcing policies to safeguard data and information from unauthorized users. Virus software, firewalls, and password control are all part of an IT professional's domain.

IT also maintains multimedia equipment for companies. Phone systems are often tied into a company's computer network. Larger companies use computers to answer and route telephone calls, process orders, update inventory, and manage accounting, purchasing, and payroll activities. Training on computer software and systems within an organization is also often the responsibility of IT personnel.

8.3 Lean Manufacturing

Lean manufacturing is a management philosophy derived from the Toyota Production System. It is focused on the elimination of waste in various forms from the production process. The Toyota Production System originally identified seven different types of waste, or *Muda*—a Japanese term meaning uselessness, futility, or wastefulness. The seven wastes describe resources that are often wasted in the manufacturing process:

1. *Transportation*: Every time a product is moved in a process it adds time. It also runs a risk of being lost, damaged, or delayed, and it does not add any value to the product that is being transported.

2. *Inventory*: Raw materials, works in progress (WIPs), and completed products represent capital outlay that is producing no income. If items are not being actively processed, they are considered to be wasting time and capital.

3. *Motion*: Excess motion of machinery contributes to wear on equipment, while excess movement of operators can contribute to repetitive stress injuries. Motion also increases the possibility of accidents that can damage equipment or injure personnel.

4. *Waiting*: If a product is not being processed, it is waiting—wasting both time and space. Most products spend most of their lifetime in a manufacturing plant waiting.

5. *Overprocessing*: Anytime more work or operations are performed on a product than is necessary, it is considered to be overprocessed. This includes tools that are more precise or expensive than required or machinery that is overly complex.

6. *Overproduction*: When more goods are produced than is required by existing orders from customers, the product is overproduced. Creating large batches of product often creates this condition since customer needs can change while product is being made. Many consider this to be the worst from of waste since it can hide and generate the other forms of Muda. Overproduction leads to excess inventory, more storage space, and additional movement of product.

7. *Defects*: Extra cost is incurred in handling the part, wasted material, rescheduling of production, and extra transportation of defects.

Lean manufacturing uses a number of tools in identifying and eliminating waste. The application of the lean methodology is often paired with that of Six Sigma and used within the management, manufacturing, and quality departments. This gives rise to the term *Lean Six Sigma* as an approach to business and manufacturing. The two systems can be used together to complement and reinforce each other as a strategy to make systems more efficient.

One of the basic tenets of the Toyota Production System and lean philosophies is that of just-in-time (JIT). This is an operation system where materials are moved through a system and are delivered with precise timing just as they are needed. This reduces in-process inventory and the associated carrying costs. To accomplish this, information about the process must be monitored carefully to ensure that product flows smoothly.

Another Japanese term used in the Toyota Production System is *Mura*, which means "unevenness" or "irregularity." Leveling production and eliminating waste through the application of the proper techniques should lead to a smooth and predictable workflow. Mura is avoided by applying JIT techniques properly.

A third Japanese term describing waste is the word *Muri*, meaning "unreasonableness" or "impossible beyond one's power." Another way to describe Muri is the overburdening of machinery or personnel. Examples of Muri are workers performing dangerous tasks or working at a pace beyond their physical limits. The concept also applies to machinery and running a system or production line beyond its designed capabilities.

8.3.1 Kanban and "Pull"

To meet the objectives of JIT, one of the techniques used is *Kanban*. This is a manual system that uses signals between different points in the manufacturing process. Kanban applies to both deliveries to the factory and to individual workstations. The signal may consist of cards or tickets that indicate the status of a bin or storage area or simply an empty bin. These act as a trigger to replace the bin with a full one and order new parts.

Electronic methods of Kanban are sometimes used within enterprise software systems. Triggers in this case may be manual or automatic. If stock of a particular component is depleted by the amount of the Kanban card, electronic signaling can be used to generate a purchase order with a predefined quantity to a vendor. Arrangements are made with the vendor to ensure that material is delivered within a defined lead time at a specified price. The result of a successful Kanban system is the delivery of a steady stream of containers of parts throughout the workday. Each container usually holds a small supply of parts or materials, and empty containers are replaced by full ones. An example of the implementation of Kanban might be a three-bin system, where one bin is on the factory floor in use, one bin is in the storeroom, and a third bin is always ready at the supplier's location. When the bin being used has been consumed, it is sent with its Kanban card to the supplier. The bin is replaced on the factory floor with the bin from the storeroom. The supplier then delivers a full bin to the factory storeroom, completing the loop.

To track Kanban cards, a visual scheduling tool called a *Heijunka Box* is sometimes used. This is a rack or series of boxes placed on a wall to hold Kanban cards. A row for each component is labeled with the name of the component or product. Columns are used to represent time intervals of production. Cards are often made in different colors to make it easier to identify the status of upcoming production runs. *Heijunka* is a Japanese term meaning production smoothing. It originates from the Toyota Production System and the concept that production flow will vary naturally and that the capacity of machinery and personnel will be forced or overburdened at certain times (Muri).

In supply chain management, one of the concepts that drives production is "push" versus "pull." In a push strategy, product is

Chapter Eight

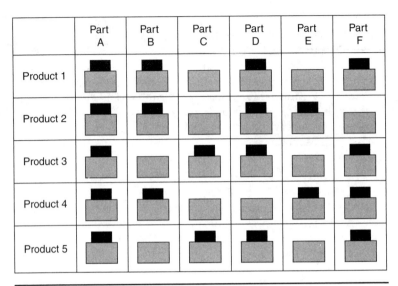

Figure 8.2 Heijunka Box.

pushed toward the consumer. Forecasting of customer demand is used to predict how much of a product will be needed. This can lead to excess or insufficient inventory since there is always inaccuracy in forecasting. In a pull strategy, production is based on consumer demand and response to specific orders. Material or parts are replaced on demand, and only what is needed is produced.

This does not mean that products must be made to order to satisfy JIT and the pull system. A limited inventory is often kept on hand or in process and is replenished as it is consumed. A Kanban system is a means to this end.

8.3.2 Kaizen

Continuous improvement, or kaizen, is a major element of the TQM system and is used in manufacturing, engineering, and business management. It involves all employees of a company, from the CEO to production workers. By improving and standardizing processes, waste is eliminated, achieving the goals of lean methodology.

Kaizen is a daily process that teaches employees to apply a scientific method to the identification and elimination of waste. It can be applied individually, but often small groups are formed to look at specific applications or work areas. The group may be guided by a line supervisor but may also be assisted or led by people trained in Six Sigma techniques.

Kaizen events are often organized as a weeklong activity to address a specific issue. They are sometimes referred to as a "Kaizen Blitz." They are generally very limited in scope and involve all personnel involved in the process. Results from a kaizen event are often used in later events after careful evaluation.

A kaizen cycle can be divided into several steps. The first is to standardize an operation or its activities—in other words, ensuring that a system is in place in the beginning. The next step is to measure the operation using whatever measurement is appropriate. Examples include cycle time, waste material, in-process inventory, or defective parts. These measurements are then compared to the design requirements of the operation. Innovations and improvements can then be applied to the process and, hopefully, improve productivity. In turn, the results of this cycle become the new standard, which is used as the basis for the next kaizen, and the cycle repeats.

Another name for this cycle is PDCA—an acronym for Plan-Do-Check-Act or Plan-Do-Check-Adjust. The planning stage is used to establish the objectives and processes necessary to achieve the desired results. Doing involves implementing the plan, executing the process, or making the product. Data collection about the process is also performed during this step. The checking stage studies the results gathered during the previous step. Comparing the results to what was expected is made easier by charting the data and making it easier to spot trends. The acting or adjusting stage is used to apply corrective actions to the process. The PDCA cycle predates kaizen by many years and has its roots in the origination of the scientific method hundreds of years ago.

8.3.3 Poka-Yoke

Safeguards that are built into a process in order to reduce the possibility of making an error are referred to in lean manufacturing by the Japanese term *poka-yoke*. This error-proofing technique is used to prevent, correct, or draw attention to mistakes as they occur.

The originator of the term was a Japanese industrial engineer named Shigeo Shingo who acted as a consultant to Toyota in the 1960s and 1970s. Shingo believed that errors were inevitable in any manufacturing process, but if mistakes were detected or prevented before products were shipped, the cost of these mistakes to the company would be reduced. Detecting a mistake as it is being made is known as a warning poka-yoke, while preventing the mistake is called a control poka-yoke.

Three types of poka-yoke are recognized as methods of detecting and preventing errors in a production system. The *contact* method tests a product's shape, color, size, weight, or other physical property. This method is often implemented in automated processes by using sensors and test stations. The *fixed-value*, or constant number, method alerts the operator if a number of movements have not been made. The *motion-step*, or sequence, method determines whether the required steps in a sequence have been followed. In automation, this method is performed by creating a fault in a system when a step is either not

FIGURE 8.3
Pick-to-light.

completed within a prescribed time or if an actuator does not complete its movement correctly.

In manual assembly, there are several techniques that are used to error-proof assemblies. Pick-to-light systems use signals to guide operators through the correct assembly process by blinking a light at the correct bin location for the component that is to be installed. Sensors are used to detect whether a part was removed from the proper location and/or installed correctly. Manual tools, such as torque drivers, can be used to ensure that the proper angle and torque are achieved when installing screws. Carefully shaped fixturing and tooling are used to prevent parts from being placed incorrectly, and embedded sensors validate placement.

Ideally, mistakes are detected as they occur. Production lines often have test stations at several points in the process to eliminate further processing of reject parts. Test devices, such as gauges, leak testers, machine vision, and weighing systems, are used to flag parts or mark them for removal. Reject bins or reject spurs are equipped with sensors or identification systems to ensure parts do not proceed through production.

8.3.4 Tools and Terms

Lean manufacturing tools have been developed over years of refinement to improve the efficiency of production. In addition to the techniques of Kanban, kaizen, and poka-yoke, other methods to aid in the organization of the workplace and analysis of data have been adopted by many manufacturers.

Standardized work instructions (SWI) allow processes to be completed in a consistent, timely, and repeatable manner. This technique involves testing work processes in order to determine the most efficient and accurate way to accomplish a task. Work instructions are created using photos, simple text, and diagrams to clearly indicate what an operator is to do. Employees often want to do things their own ways, but developing the most appropriate method to accomplish a task is a critical element of process improvement. Workers should be encouraged to challenge the instructions and help make improvements, but consistency is reinforced when everyone is completing tasks in the "current best way."

The 5S system creates a workplace that is clean, organized, and free of materials not needed for production. The 5S system, sometimes called the "visual workplace" or "visual factory," is a housekeeping tool with five behaviors intended to make the workplace more effective.

1. *Sort*: Decide which items are needed to accomplish the task and remove all others.
2. *Straighten/Set in Order*: Organize the items in a work area so that they can be accessed quickly. Make a place for everything, and put everything in its place.
3. *Sweep/Shine*: Clean and inspect everything in the work area. Perform equipment and tool maintenance regularly.
4. *Standardize*: Use standard procedures and instructions for all work. Use discipline and structure to maintain consistency.
5. *Sustain/Self-Discipline*: Continue to maintain the 5S efforts through auditing and documentation. Ensure that employees understand company expectations and the need for an uncluttered and organized workplace.

Value stream mapping (VSM) is used to design and analyze the flow of products or information between key work processes. This technique is used to differentiate value adding activities from non-value-adding ones and reduce waste. Software tools and templates are available for accomplishing this task, but VSM is often performed by teams drawing diagrams by hand. Shigeo Shingo suggested that value-added steps be drawn horizontally across the center of a page and non-value-added steps represented by vertical lines at right angles to the value stream. Shingo referred to the value-added steps as the process and the "waste" steps as operations. Separating the steps allows them to be evaluated using different methods; VSM often leads to the discovery of hidden waste activities.

OEE is used to monitor and improve the effectiveness of manufacturing processes. It is described in terms of its application to machinery in Sec. 2.9.

Statistical process control (SPC) refers to the application of statistical tools to the monitoring and control of a process in order to ensure it operates at full potential. The goal of a process is generally to produce as much conforming product as possible with a minimum of waste. Inspection methods detect defective product after it is made, while SPC emphasizes early detection and prevention of problems. SPC is not only used to detect waste; it can reduce the time that it takes to produce a product. It is used to identify bottlenecks in a process, waiting times, and sources of delay. SPC can be used on any process where the output of conforming product (product meeting specifications) can be measured.

SPC monitors a process by using *control charts*. Control charts use objective criteria to distinguish background variation ("noise") from significant events. The first step in SPC is to map the process, breaking it down into individual steps. This can be done using a flowchart or a list of subprocesses. Typical variables identified during the process mapping step include downtime, defects, delays, and cost. The next step is to measure sources of variation using control charts. A common type of control chart is a line chart. The line chart is used to correlate measurements over time or a number of samples.

By adding an average line to the line chart, a *run chart* is generated. A run chart shows the variations of the process from the average over time. Adding upper control limits (UCL) and lower control limits (LCL) creates a control chart.

The type of data that is being collected determines the type of control chart to use. Attribute control charts include p charts, np charts, c charts, and u charts. Types of variable control charts include Xbar-R charts, Xbar-s charts, moving average and moving range charts, individual charts, and run charts. Descriptions and further treatment of these can be found in SPC books or online.

Control charts provide a graphical method of viewing when a process exceeds control limits, but there are many rules that must be applied to the statistical results to determine whether a process is stable or not. If none of the various detection triggers occurs, the process is determined to be stable. If a process is found to be unstable other tools, such as *Ishikawa diagrams*, designed experiments, and *Pareto charts*, can be used to identify sources of excessive variation. Ishikawa diagrams, also known as fishbone or herringbone diagrams, are line drawings that show the causes of a

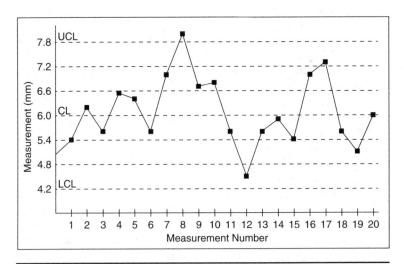

Figure 8.4 Control chart.

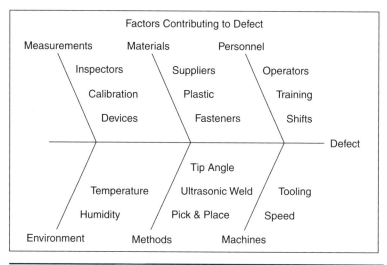

FIGURE 8.5 Ishikawa or "fishbone" diagram.

specific event. Causes are usually grouped into categories such as people, methods, machines, materials, measurement, and environment, which are then further subdivided into potential factors contributing to error.

8.4 Systemization

The information and discussion in this section of the chapter is solely the opinion of the author after observation of the working methods used in various companies.

The tools discussed in this chapter are just that: tools. Like any tool they can be misapplied and misused. Tools do not perform any work themselves; they have to be applied by people. Therein lies the problem: everyone does things differently. One of the key elements of implementing a successful business, project, or machine is systemization. This can be described as a method of ensuring that things are done the same "current best way" every time.

With all the tools and resources available, it is sometimes easy to arrive at "analysis paralysis," where a great deal of time is spent determining which solution or solutions to apply to a problem, but no solution is ever applied. A systematic method of choosing a solution involves careful analysis of the problem, establishing a *deadline* for selecting a method, and application of the solution based on a criteria such as most appropriate, least time-consuming, lowest cost, simplest, and so on. It is important not to implement multiple solutions that may be at cross-purposes with each other. Many lean manufacturing and Six Sigma experts are very experienced in selecting the appropriate solution; bringing in outside help for defining and solving company problems can be a wise investment.

Even if the correct tools are used, incorrect inputs can lead to errors in output. There is an acronym for this used in the computer industry, GIGO, or "garbage in, garbage out." This not only applies to programs and algorithms; it also applies in the business world. As an example, when discussing a problem with involved personnel, their opinion can be colored by relationships with other people also involved. An unbiased opinion can be hard to obtain, and direct observation of operations is usually the preferred method. Ideally all information used as input to solve a problem should be quantifiable.

8.4.1 Job and Task Descriptions

The SWI described in the previous section apply to processes typically found in the manufacturing environment. They are often represented by pictures of activities and numbered, written descriptions of how a specific task is to be done. This same method can be applied to nearly any job function in the workplace, although it may not be necessary to post them in a public location as you might see on a factory floor.

All too often the most basic of job descriptions do not exist in many businesses. Employees are hired as "manufacturing engineers," "buyers," "sales," or "maintenance," as if these functions are the same in every business. This can lead to disappointment and frustration when jobs are not performed as expected.

Ideally job and job function descriptions should be provided to employees before the first day of work. This provides a list of expectations and general functional descriptions for each position in a company. This should be a "living" document that is updated on a regular basis as job functions evolve, which they generally do over time. The document should start with a general overall description of the position, the title of the person filling the position, where in the overall management hierarchy the position falls, and the names of employees that relate with the person in this position on a regular basis. This gives new employees a sense of belonging within the overall organization and within their department.

After the general description of the position, an outline of the functions within that job should be listed. This can be broken down into as many levels as necessary; again this should be an evolving document. The statement "it's not my job" should never occur within an organization that has a properly documented list of job function descriptions. As conflicts appear as to where a particular function falls within an organizational map, the issue should be documented for discussion at the next job function meeting. Of course, until the conflict is resolved someone still has to perform the task.

Job function meetings should be held at least on a monthly basis initially. They can be informal and last as little time as necessary, but they should never be put off or ignored. Input should be gathered from every person possibly affected by a change in function; members attending the meeting should include at a minimum immediate supervisors of the position being discussed; department heads should also be involved if possible.

Feedback from employees should be a part of the process. In the initial stages of documenting a job and its functions, it may take several kaizen-type meetings to produce a satisfactory initial document. As the document is fleshed out, the process can be done on a more periodic basis and with less involvement by nonaffected personnel. Since the description falls under the "current best way" category, efforts should be made to keep the document as accurate as possible.

8.4.2 Communications

As part of daily activities in the workplace people engage in many different forms of communication. Verbal, electronic, printed, or handwritten and even nonverbal methods are commonly used. Choice of the proper method of communications is a critical part of the business process.

Meetings are a major component of most business methods. They are intended to be a method of two-way communications between individuals or groups of people. All too often they end up being more of a one-way method, where a presenter provides information to a group of people without much feedback. This type of meeting is often more appropriately done through a written method for two important reasons. The first reason is that meetings take more time than reading a presentation. They are also less flexible in terms of timing; they are scheduled for a specific time during which everyone must stop what they are doing and attend. So the first disadvantage is that of time.

The second reason applies to more than just meetings: verbal presentations are often not documented. Sometimes an outline is presented to attendees; people may even take notes. The problem here is that there is no guarantee that the important information was received and understood by all attendees in the same way. This reemphasizes the importance of clear, written communications in business.

Meetings can serve an important function: when two-way communications are needed between individuals or groups, they are the most efficient way to exchange information or come to an agreement. Even here it is important to document them with written communications so that there is not later disagreement over what actually took place. Most people in larger organizations would agree that too much of their time is taken up by meetings. One way to reduce the number of meetings is to determine whether they can be replaced with written forms of communication.

Conversations between individuals should also often be documented by written communications. If an important decision or instruction is not documented, there may be disagreements later as to what was actually said. The statements "you never told me" or "that's not what I/you said" should never be made if proper communication methods are chosen. With the development of technologies such as personal communication devices able to send texts or e-mails, there is simply no excuse for miscommunication or misinterpretation. Especially for project-based work, documented communication between coworkers and vendors/customers is critical.

8.4.3 Hiring and Training

The hiring process often includes rigorous analysis of a prospective employee's experience and skill set, suitability for a position, and ability to "fit in" with an organization. Candidates are often interviewed by multiple people, including (hopefully) peers familiar with the person's field of expertise. If an employee is being hired for a position that requires specific skills, testing of the candidate should be part of the process. Examples in the technical field might be generating CAD drawings, writing a short program, fabricating a part, or troubleshooting a machine. Decisions on whether a candidate for a position is to be hired are usually based on various factors, including personality, experience, skills, and compensation requirements.

Even with all this careful examination of a candidate, there is a variety of skills that will likely not match exactly. Every company has a different method of accomplishing tasks, and new employees often need guidelines or training to completely mesh with their new organization.

If the job and task descriptions described in Sec. 8.4.1 are detailed enough they can serve as an important training tool. Examples of previous work or projects can also be a useful element in the training process. If an employee with a similar job function is available, he or she may also be assigned to help with the orientation process.

Even with all these internal resources, sometimes outside help must be obtained to facilitate the training process. Vendors such as distributors' and manufacturers' representatives can be an excellent resource for product-specific training. They will often provide "lunch and learn" activities in order to help promote their products. Trade shows also often provide hands-on training on many products.

Skills such as CAD/CAM and lean manufacturing techniques are often taught by independent companies and consultants. Establishing a budget for employee training can be a wise investment for a company provided employee retention is high. Training programs are often administered by the human resources department.

A common problem is balancing training with retention of employees. Companies often hire personnel at a relatively low pay rate with the intent of training them into a position, only to find that once the employee's skills have increased, he or she has become more marketable and goes to find a higher-paying job. It is important that companies stay competitive in the job market while still controlling their own bottom line.

8.4.4 Engineering and Project Notebooks

An important tool often used by inventors, engineers, and designers is the engineering or project notebook. The intent is to capture vital details of the engineering process and create an ongoing record of the project. Observations, ideas, and even meeting notes help provide a timeline for the project. Formal engineering notebooks are often used as a permanent record of a project accessible by all members; these are usually bound books (so that all pages are accounted for), written in ink (pencil and erasable ink are not acceptable), with all entries dated and signed. There are often formal methods of making corrections (no Wite-Out, a single line drawn through errors), and they may even contain signatures of team members who attended project meetings.

While formal engineering notebooks may be used in large companies and engineering firms, they are not common in factories and industrial plants. Perhaps people believe that they take too much time to maintain or they have simply never heard of them.

I have created a notebook for every project I have ever been involved with. While breaking nearly every rule listed above (mine are usually written in pencil, I use loose-leaf binders, and I do not sign anything or expect others to), I still find them very useful. I have often gone back to a project notebook to review how I solved a particular problem or wrote a piece of code.

In addition, I use a different notebook to keep records and ideas in that do not fall into a particular project. My project notebooks are always available for other team members or employees to look at, but my personal notebook contains scheduling information, random thoughts, and even passwords for the various web sites and unimportant accounts I am signed up for. In some ways it is a hybrid journal and life-project notebook.

As mentioned earlier in this section, I am a big believer in written communications and records. Many times I have visited a plant to evaluate a project only to find that there is insufficient information on a machine or system to be able to properly estimate the time it would take to solve the problem. Documentation is the primary and most important tool used in troubleshooting. This not only applies to the formal documentation that is supplied with a manufactured system, but also to the maintenance records and notes that operators and technicians keep on the equipment. For the cost of some paper and a little bit of time, it is possible to save thousands of dollars by simply recording the events in a system's history.

This can also be extended to the business world. By keeping written documentation of daily events, it is much less likely that tasks will be forgotten. There are various electronic devices such as personal data assistants (PDAs), laptops, iPads, or even cell phones that can aid in this task, but I personally still prefer the handwritten method.

CHAPTER 9

Machine and System Design

This chapter describes the typical cycle of machine quoting, procurement, design, fabrication, debug, and installation.

After a recent improvement in the business forecast, the ACME Widget Company has discovered that they will need a new widget line within the year. Mark, a company project engineer, has been tasked with finding appropriate vendors for the equipment and getting the line installed and started up for production.

9.1 Requirements

Mark's first task is to define the requirements for the new production line. A meeting is held with the engineering manager, production supervisor, quality manager, maintenance manager, and process manager. The company's vice president of operations and comptroller are also present.

The forecast for the next few years' production is that about 100,000 widgets per year will be required for the first couple of years ramping up to between 150,000 and 200,000 per year for the next few years after that. In addition, a slightly larger widget has been designed that will be put into production next year. This brings the total number of different kinds of widgets produced by ACME up to seven.

9.1.1 Speed

ACME currently works two eight-hour shifts a day. The current two production lines produce about 70,000 widgets per year, the bulk coming off a newer line installed about five years ago. This line produced about 42,000 widgets last year but was down for maintenance, repair, or setup/changeover about 20 percent of the available time.

The total days worked excluding weekends, holidays, and schedule shutdowns are 48 weeks multiplied by 5 days per week, or 240 days per year. This comes out to 3840 hours per year of available production time. If the total downtime of 768 hours is subtracted, this leaves 3072 hours of actual production time. Since 42,000 widgets came off line B in 3072 hours, this means the line averages about 13.67 widgets per hour.

The production supervisor says he is sure that the line was specified at 16 widgets an hour and had achieved 18 during the initial runoff. After some quick calculations, the quality manager determines that the line is running at about 85 percent efficiency (13.67/16 x 100% = 85%), discounting maintenance, repair, and changeover. Since the ACME company has an active Six Sigma/lean manufacturing program, the quality manager notes that this would be a good subject for a kaizen improvement project to reduce downtime.

After more discussion, it is decided that the specification for the line speed will be set at 18 widgets per hour. If the line efficiency can be raised to 90 percent, this should allow for (3840 hours x 18 widgets/hour) x .9 efficiency = 62,208 widgets. This should easily accomplish the initial target of 100,000 per year but will leave the capacity short after about 1.5 to 2 years if the forecast is accurate.

9.1.2   Improvements

The production supervisor and maintenance manager believe that if some of the current pneumatic actuators were replaced with servos setup time would be reduced by up to 50 percent. Since setup time is estimated to be about 40 percent of the total downtime, this could be a significant improvement. In addition, several problem areas on the current line seem to cause most of the downtime.

The quality manager asks if data is collected to help determine the cause of line stoppages. The production supervisor states that data had been collected during several periods over the lifetime of the production line but not recently because of manpower constraints.

The engineering manager mentions that it would not be terribly expensive to implement a data collection system using the existing PLC on the line. In addition, the information could be collected into a production PC located on the factory floor.

Mark has been noting all these comments during the meeting. He asks the VP of operations and comptroller if a budget has been determined for the project.

9.1.3   Cost

The comptroller has a report from the implementation of the line five years ago. The line had taken a full year to build and install from start to finish. It had cost in the neighborhood of $2 million, including facilities improvements, in-house labor, and payments to vendors. At that time there had been a much larger engineering department and much of the system design and layout had been done internally, including implementation of the packaging system.

The VP of operations says that he thought they would have about $2.4 million available for this project, but that Mark would not have much access to internal labor. He mentions that he thought he could think of a few areas where costs could be reduced and would get together with Mark and the engineering manager the next day to discuss some of these ideas.

9.1.4 Requirements Documentation

Mark uses the company template to begin generating a requirements document for the new widget line. Many of the items, such as electrical and mechanical specifications, have already been included in the template, so Mark starts by making a list of the requirements specific to this project. Information such as the line speed capability, packaging requirements, footprint or space requirement, and dimensions of the widget, including the new product, are included. Much of this information has to be collected from different departments within the company.

Mark also knows that per company requirements he will need to get a minimum of three quotes from machine builders and integrators. Because of this, the requirements documentation needs to be very complete to minimize questions.

9.2 Quoting

Mark knows of several machine builders both locally and nationally that are appropriate for building the new widget line. The company that built line B five years ago is from pretty far away but had done a decent job on the previous line. They were very large, however, and not always as responsive as he would have liked.

Another company had recently made a few sales calls on both Mark and the engineering manager, trying to drum up new business. They were from the same city as ACME and had left some brochures with Mark describing some of the jobs they had done. The brochure looked very professional and had some nice pictures of some pretty impressive-looking machines, but nothing looked very similar to what this line would look like.

There are two other large machine builder/integrators that Mark knows can build the line, again both very far away. After going online and checking out their web sites, Mark decides to send quote requests to LineX, the company that had built line B; LocalTech, the company that had called on them recently; and the Mammoth Corp, one of the two large national machine builder/integration companies.

9.2.1 Quote Request

Mark writes up a fairly brief quote request with a description of the proposed line, a picture and description of the widget, and a statement that further information, details, and specifications would be sent if the company would be interested in quoting the line. He e-mails the request to all three companies and begins further refining the requirements document.

Later the same day, Mark receives a call from Bill, the applications and sales engineer from LocalTech. Bill asks if he could stop in the next day and take a look at the existing Line B. Mark replies that tomorrow would be fine and that he would have the requirements finished by then so that Bill could take them with him.

The next morning Mark receives another call from Jack, the sales rep for LineX. Jack says that he will be in town early next week and he would like to stop in and meet with Mark at his convenience. Mark says sure and puts Jack on the schedule.

When Bill arrives, Mark has him fill out a standard nondisclosure agreement with the receptionist. He then takes him out onto the plant floor, first picking up a set of safety glasses and earplugs. Mark and Bill spend about two hours watching the line run and discussing the operations. Bill asks several questions about hardware, sequencing, and when the order might be expected to be placed. Bill also takes quite a few notes as he examines the line. Mark gives him a copy of the requirements documentation, including a machine specification. Bill says it will probably be the end of next week before he can work up a quotation. Mark also gives Bill a sample of a widget directly off the line for Bill to take with him.

The following week Jack arrives. Mark discusses the previous line's operation and mentions the slow response of LineX to several machine issues over the past few years. Jack explains that LineX has undergone a reorganization and that they now have a separate service department and are much more responsive. Mark gives Jack the requirements documentation and a sample widget to aid in his quoting process.

Since Mark has not heard anything from the Mammoth Corp, he gives their sales department a call. He is connected to the regional applications engineer, Steve, who explains that they have been inundated with quote requests and would not be able to make it out to look at the application for a couple of weeks. They set up a meeting for Thursday a couple of weeks later.

By the time Steve makes it out to ACME, Mark has already received quotations from the other two companies. Steve says he should be able to turn a quote around in a week or so, but that deliveries are running a bit long currently.

9.2.2 Quote Analysis

Two weeks later a meeting is set up with the same members as the kickoff meeting to discuss the quotations. The Mammoth Corp barely got their quote in on the evening before the meeting, making Mark worry about the comment Steve had made concerning long delivery.

The Mammoth Corp provided the most detailed and comprehensive of the three quotes. Much of it appeared to be boilerplate, but it was obvious that they had done their homework. It was accompanied by a nice cover letter and 3-D renderings of what the finished system would look like. It was also the most expensive of the three quotations at about $2.2 million.

LineX provided a detailed quote also with pictures of line B and detailed descriptions of machine operation. Improvements to the new line were also described, including an innovative part-flipping mechanism. The quote was the lowest price of the three companies at a bit over $1.7 million. Since the previous line had cost almost $1.6 million, this seemed quite reasonable with the proposed improvements. Of course, having built the previous line, it was expected that LineX would be the least expensive of the three.

LocalTech's quotation came in at just under $2 million. Like the Mammoth Corp, the quotation included 3-D solid models of the line. Details were also broken out into sections proposing improvements to both the original line and the new one.

It was obvious that LocalTech was very hungry for the job; emphasis on the local support was evident in the quotation. Assurance that they had the expertise required for the application was also written into the proposal.

9.2.3 The Decision

After weighing all the factors, it was decided that a purchase order would be issued to LocalTech. Although the pricing was higher than that of LineX, it was agreed that the local support could be valuable in the long run. Mark had called several of LocalTech's references and it was generally agreed that LocalTech stood behind their work and really went the extra mile for the customer.

Mark drafted a *statement of work* (SOW) that effectively restated the items in the quote request and set out the tasks and expectations of both the vendor and customer.

9.3 Procurement

Mark called Bill from LocalTech with the good news that they had decided to use them to build the new line. Bill stopped by later that afternoon with his engineering manager, Jim, to discuss the procedures that would be used to manage the project. During the discussion, the conversation got around to payment terms and Bill mentioned that there might be a bit of a problem....

9.3.1 Terms

LocalTech's requested terms were 40 percent down payment with order, 40 percent prior to shipment, and 20 percent net 60 days after delivery, or 40/40/20. ACME had a standard policy of 30 percent down, 30 percent after receipt of 90 percent of materials, 30 percent after FAT and prior to shipment, and 10 percent net 90 days after successful SAT. After a phone meeting was held between ACME's buyer and the owner of LocalTech, it was agreed to abide by ACME's terms. A clause was written into the purchase order that the second 30 percent would be paid no longer than 120 days after initial parts requisitioning by LocalTech to help with cash flow.

A purchase order was faxed to LocalTech two days after the meeting and mailed one day later. LocalTech sent an invoice to ACME, and payment was received 57 days after that.

9.4 Design

A project kickoff meeting was held at LocalTech to discuss the application. Bill and Jim described the project, and the quotation was presented to the project team. LocalTech had two project managers that managed the budget and schedule for machines and systems; Paul was assigned to manage this project. He reviewed the quotation documents and generated a Gantt chart using Microsoft Project for distribution to the project members.

9.4.1 Mechanical

Joe was assigned as the mechanical lead on the project. There were three mechanical project engineers at LocalTech and it was fortunate that Joe's previous project was winding down since he had built a line very similar at his previous job. Joe asked if any documentation from the previous line B would be available to leverage for the project. Bill said he did not think that Mark from ACME would feel comfortable giving them any solid models or CAD files since LineX had bid on the line but had not gotten the job. Everyone agreed that that would probably not be an ethical thing to do.

The first thing Joe had to do, however, was read the specifications. After asking a few questions about hardware choice restrictions, he was ready to start work on the design.

Joe first generated a timing and device chart using his template. This was a Microsoft Excel template that was used by applications, mechanical, controls, and software departments jointly to determine machine timing, pneumatic sizing, and the actuators to be used on the machine.

The next step in the design process was to start designing machine assemblies. An assembly is a grouping of components to accomplish a specific task. It may consist of a simple actuator or something as complex as a multiaxis pick-and-place with mixed electrical and pneumatic cylinders. Though this could be decided arbitrarily by the mechanical designer, LocalTech had a standard procedure for deciding what would make up an assembly.

Joe decided that a Stelron chassis would be used for the movement of widgets around the machine. Widgets would be fed from a vibratory feeder or bowl onto a pallet on one corner of the chassis via a two-axis pneumatic pick-and-place with gripper. After proceeding clockwise around the chassis through various assembly and inspection stations, widgets would be removed at the adjacent corner after moving across three sides of the chassis. Empty pallets then rotated around to be loaded again.

Since the process would be mostly synchronous, some of the actuators would be moved by the chassis itself. Those that were not had to be dimensioned and sized to determine their capability and suitability.

Mechanical Drawing

All LocalTech's designs were done in SolidWorks, a 3-D solid modeling program. Joe had worked on this platform for over 10 years and was quite comfortable with it; in fact he had generated the solid model representations of the line concept for Bill.

Joe decided to spend a day or so downloading files from vendors for the chassis, actuators, and other purchased components that would be used on the system. There were four additional mechanical designers at LocalTech for a total of seven mechanical licenses, or "seats." The designers acted as a resource pool for the mechanical department, filling in where necessary for the project engineers. After Joe had spent some time assembling a list of proposed components (preliminary bill of materials), he gave part of the list to a designer and downloaded the rest himself.

Most of the vendors for the major components had step files, a generic file format that worked across solid modeling platforms, available for their hardware. A few had only AutoCAD files in three views; these would have to be turned into solid models.

With the help of the mechanical designer, it took about three weeks to get all the individual assemblies modeled.

Finite Element and Stress Analysis

After drawing all the assemblies and floating them above the chassis locations, Joe entered the estimated weights and dimensions of hardware into a spreadsheet and did some analysis on the expected stresses that would be incurred by the mechanisms, bracketry, and framing of the main machine. Joe had a finite element method (FEM) software package that allowed detailed visualization of where structures would bend or twist. This allowed him to produce stiffness and strength calculations that would help minimize and optimize weight, materials, and costs throughout the machine.

Framing and Bracketry

With the FEM analysis done, the mechanical team was able to attach all the assembly models to the chassis. As with any design, various interference points were found and brackets and locations had to be modified. One of the advantages of using a chassis was that locations were fixed; if there was not enough space for an assembly, one could simply ensure that the required space was available by leaving empty pallet stations. In one case a mechanically linked station had to be moved over one spot; the chassis was requoted by Stelron with no change in price.

Joe decided that most of the framing and brackets would be made of welded cold-rolled steel. Of course, this had to be painted, but it was typically the least expensive option. An exception was the mounting for the vision system: Joe decided to use aluminum extrusion since the camera needed to be adjustable. A few of the sensor mounts were made of aluminum plate since load requirements were not high, the brackets were small, and they did not need to be painted.

Joe also decided to use aluminum extrusion for the guarding on the upper frame of the machine. This way it would be easy to make doors and mount hinges and switches; also it was much easier to correct mistakes that might be overlooked during the design process.

Detailing

After a mechanical design review with the customer, Joe and his mechanical designer started detailing the individual components of the machine for manufacturing. This involved breaking the assemblies down into their individual parts again and dimensioning, tolerancing, and specifying finish and materials on two-dimensional drawings. Some of this data could be put into a CNC machine directly, but much of it would be made the old-fashioned way by machinists on mills and lathes.

Much of the hardware was detailed here also; bolt and screw sizes and plate thicknesses were determined to generate the final BOM. Plate and tubing thicknesses sometimes had to be ordered oversize because of the skimming of material during the machining process.

Instructions for the machinists were also placed on the detail drawings, things like which pieces to stress relieve and whether assemblies had to be machined again after welding. This was especially important for some of the machine pads that needed to hold brackets completely parallel.

9.4.2 Electrical and Controls

Gordon had been at the initial project kickoff meeting and had also read the specifications. Most of his controls hardware brands were pretty well spelled out in the specifications and requirements documentation, so nothing looked particularly difficult. ACME was pretty heavily specced on Allen-Bradley controls. The original controller had been an SLC 5/05. Most of what was being used in the industrial world nowadays was of the ControlLogix family. Gordon

Machine and System Design

gave Mark from ACME a call and asked if it would be OK to use ControlLogix.

Mark did not know offhand, but after a quick call to Jake, the controls tech on the plant floor, he had his answer. ACME did have Allen-Bradley ControlLogix processors in the plant, but since they had not renewed their license, they were about two revisions back on their software. Gordon assured him that would be OK since he had revision 17 on his laptop.

Gordon had met with Joe early in the design process to help him tweak the timing and device chart. They had agreed on the names they would use for the actuators and sensors, and Joe had done a lot of the sizing of the pneumatic cylinders already. The first thing Gordon started working on was the I/O list, which would dictate the layout of the controls.

One of the improvements LocalTech had proposed was using networked I/O on the machine to save space. They had used DeviceNet and ControlNet on quite a few projects, but Ethernet I/P-based I/O was very cost-effective and would be used on this system.

Since there were a couple of external systems, including a bowl feeder, vision system, and two packaging machines, to be integrated, networking and distributing the I/O made a lot of sense. The vision system was to be a Cognex Insight smart camera, which already had Ethernet I/P connectivity and a host of examples on their web site. There was no controller on the vibratory bowl, but the sensors were to be paralleled into a local node for data acquisition.

After calling the vendors for the two packaging machines, it was discovered that one was an Allen-Bradley SLC 5/05, so Ethernet I/P would be no problem. The other was a Siemens S7 Controller with PROFIBUS capability but no Ethernet I/P. The interface between the controllers could be digitally "bit-banged," where inputs of one controller are connected to outputs from the other and vice versa, or a ProfiBus card was available from ProSoft for the ControlLogix rack. After weighing the cost, speed requirements, and integration time into the equation, Gordon decided on the digital "bit-bang" approach.

Gordon finished his I/O list and started a simple single-line diagram to show power supplied to the various components of the system. Since there were a couple of 480VAC motors on the conveyors and packaging systems, he drew a three-phase 480V line and placed a box indicating each conveyor and the packaging machine underneath. There were quite a few 120VAC devices, including the PLC, vibratory bowl, DC power supply, and a computer utility port, so he decided to use a single-phase transformer to supply this power internally.

For safety all sensors were to be 24VDC so he placed a single line of 24 V power under the 120VAC row. For all the boxes placed under these lines, he placed approximate fuse sizes to get an idea of the total current draw of the system.

Like Joe, Gordon had a couple of electrical designers he used to generate his actual electrical drawings. After a meeting where he provided a designer his I/O list and single lines and described the system, he began to consider his programming task.

After a couple of weeks the designer had generated a fairly complete set of drawings for review. Gordon used a red pen to "redline" corrections onto the drawings, the designer made the changes, and the drawing package was ready to review with the customer.

9.4.3 Software and Integration

LocalTech was a small company, so controls project engineers did double duty on electrical design and software programming. This of course kept Gordon and other controls engineers quite busy, but everyone took up slack for each other to make schedule.

Flowcharting

While the designer was drawing schematics, Gordon started making flowcharts for his programming. As with most of the design functions at LocalTech, there was a template for this. Microsoft Visio was used to draw logical statements and flow similarly to any computer program. Although PLCs did not operate in quite as linear a fashion as the usual Fortran, Basic, or C programs since scanning was involved, the techniques used were much the same.

Automatic Code Generation

Another time-saving technique LocalTech used in program design was a spreadsheet-based tool. On the I/O spreadsheet where all the I/O assignments were made, there were a number of Excel macros written to generate AutoCAD descriptions for the designers and tags for the PLC and HMI. In addition to the I/Os, a column was dedicated to defining what kind of device the I/O point belonged to. Photo-eyes and proxes, push buttons, solenoids, motors, and even servo components were designated, and internal program permissives, HMI buttons and indicators, and even fault tags were generated.

LocalTech could program quite a few different types of controller, both PLC based and others, but one of the advantages of using Allen-Bradley's ControlLogix platform was that the usually graphical ladder logic could be programmed mnemonically. This meant that text statements could be used to generate an L5K program that could then be viewed graphically. The macros in the program would generate actual ladder logic rungs with tags in the correct locations and subroutines. This saved a tremendous amount of time since many of the I/O rung subroutines as well as faults and data acquisition/OEE subroutines and rungs were generated automatically. A generic PLC program template was then opened and all these routines copied into it.

Coding

Auto sequence subroutines and other nonstandardized logic still had to be created the old-fashioned way, but the flowcharts were quite helpful there. It always helps to plan the code out structurally first. By the time the electrical design review was held with the customer, Gordon had made a good start on his coding.

Gordon preferred to use numerical values for sequence states. Bit of DINT and State Logic techniques were often used by his peers at other machine builders and integrators, but since Gordon had been at LocalTech the longest of all the controls engineers, he was able to institute this as a standard.

Each sequence was contained in its own subroutine called by the station routine. The line was logically divided into zones and stations agreed on with his mechanical counterpart, Joe.

The HMI

HMI tags were also generated by the spreadsheet and imported into the program software. An Allen-Bradley Panelview Plus HMI was used for the ACME line; sometimes customers wanted the extra capability of an actual computer for the system, but it was not required for this application.

By the time Gordon got to the HMI part of his programming and design, Joe had created a number of renderings of the entire line and some of its stations. Joe was able to easily export these as .dxf files. Gordon opened them in his AutoCAD program, then re-exported them as bitmaps to Microsoft Paint so that he could simplify them. He then imported his bitmaps into the HMI program.

As with the PLC program, there was an HMI template. It had a main screen; a couple of sample station screens; screens representing PLC I/O, OEE, and data production screens; standard servo screens; and faceplates. This gave the programmer a good jump on creating the application.

Integration

There were several external systems to the main line. The vibratory bowl fed a conveyor, which fed the chassis; the chassis fed its output to a conveyor, which passed through a packaging system including a case erector and filler. The Cognex vision system also had to be tied into the system and triggered and timed with a reject mechanism for failed widgets.

Meetings had been held with the packaging machine and vibratory bowl manufacturers when purchase orders had been placed by LocalTech. Gordon had received quite a bit of preliminary information from these vendors, but final drawings would not be completed until shortly before delivery.

LocalTech had a close relationship with a local machine vision consultant who specialized in Cognex. An order had been placed based on the number of hours Bill had thought the project might take to be set up. Gordon also had experience with Cognex vision, so after the system was initially set up, he could take responsibility and maintain it.

9.5 Fabrication

As parts started arriving from the vendors after being ordered by the design teams, they were stored on shelves next to the assigned assembly area. At any given time LocalTech generally had several projects under way, and it was important to place parts and assemblies in a common location so they could be easily located.

9.5.1 Structural

Much of the main structure of the line was supported by the chassis and conveyors, but a welded frame was needed at the two ends of the chassis to mount a pick-and-place assembly. The frames were welded together in an area off the machine shop and then finished in the CNC machine.

Most of the brackets that provided station structure were done in the same way. The frame was then stress relieved using a vibratory stress relief unit, and the bracketry was done in an oven. The pieces were then masked and sent out to be painted since LocalTech did not have a paint shop. When they arrived back at the receiving dock, they were immediately tagged and brought to assembly.

9.5.2 Mechanical

Many of the other mechanical assemblies were simply made in the machine shop, and most were to be bolted together. As with the structural pieces, when they were completed, they were tagged and brought out to the assembly area.

Several pieces were very tightly toleranced and had to be verified on the new CMM. In tolerance parts were brought out to the floor, while out of tolerance parts were either reworked or remade.

9.5.3 Electrical

After the schematic drawings had final approval from Gordon and ACME, the panel shop was given the released drawings. Components had been arriving for more than a week, but there were still not enough to warrant starting the panel-building process. Gordon gave the vendor a call and found that they were still waiting for a connector for a servo drive before they could ship complete. Gordon knew the electrical guys were eager to get started, so he told the controls vendor to go ahead and ship what he had.

Since most of the parts had arrived, the backplane was laid on a couple of sawhorses and a pencil was used to carefully mark where the DIN rail, wireway, and major components would be mounted. Holes were drilled and tapped for components and through drilled for the rivet-mounted wireway. All the major elements were then mounted to the painted steel backplane to be wired.

LocalTech had an employee who had done all the panel wiring since the company had started. There were several electricians who wired and plumbed machines, but all panel wiring was left to Mieko since her wiring was beautiful to behold. Since she had small fingers and an eye for symmetry and distance, she was well known as a fastidious and fast worker.

Larger AC wiring was used for all the 480VAC elements. This was color-coded brown, orange, and yellow so that phases could be identified easily. AC wiring was then done in red and white and DC wiring in blue and white-striped blue wire. All wiring was routed through wireway and made neat 90° turns, adhesive wraparound labels were consistently located about 1/8 in from terminal blocks, and care was taken to ensure different voltage levels were not routed together.

9.5.4 Assembly

The chassis had been on the assembly floor for a couple of weeks before the frames and structural components were delivered from the paint shop. The assembly crew had put together several of the subassemblies and stations on worktables, and now they could start putting together the whole machine.

The electrical enclosure was a low double door cabinet that mounted to the frame at the entry to the chassis. Holes were punched in the cabinet before mounting it to the frame, and the fully wired backplane was bolted inside.

After all the major stations and assemblies were mounted to the frame and chassis, the guarding structure was attached to the top of the machine. Lexan panels and doors were inserted, and much of the main chassis was complete.

The conveyors had arrived at about the same time as the vibratory bowl and chassis, but the packaging machines were running a bit late on delivery. Gordon and Joe had attended runoffs at both manufacturers and had found a number of items that did not meet ACME's specifications. While the manufacturers had agreed to fix most of the issues, a few would have to be handled by LocalTech.

Conveyors and auxiliary machinery were placed into position, and a laser transit was used to locate and level components to each other. Machinery was then lagged to the floor and the rest of the guarding put in place tying the production line elements together.

The electricians and Mieko began wiring the sensors, valve banks, and other components back to the main panel. Plastic wireway was mounted to the framework of the machine and guarding to provide convenient routing for cables and hoses. Quick disconnect cables were used in various spots, so that the machine could be easily disassembled and shipped later.

9.6 Start-Up and Debug

The original Gantt chart Paul had created showed that the project was about two weeks behind schedule. After discussing the options, it was decided that the team would try and make up the time during the start-up phase by working some later hours and Saturdays. This was a fairly common occurrence in the machine-building industry, and several of the team members appreciated the extra hours on their paychecks.

Assembly was finally ready to turn the machine over for start-up, and the project team was assembled on the assembly floor for a safety briefing. Gordon and Joe handed out a list of potential safety hazards and a start-up checklist to each of the team members. This reiterated lockout/tagout procedures and stated that only Gordon and Joe were authorized to power up and operate the machine until debug was complete.

9.6.1 Mechanical and Pneumatics

Before applying power to the machine air was applied to the pneumatic quick disconnect. A technician went to all the actuator valves and ensured that they were plumbed correctly by manually actuating the valve with a screwdriver. Flow controls were set to ensure that cylinders operated at the appropriate speed with Joe's supervision. They were then locked down to ensure they would not be moved inadvertently.

Gordon did a basic electrical check to make sure that there were no short circuits phase to phase, phase to ground, or + to -. Fuses were all checked against the schematics, and the fuse holders and breakers were left open. Gordon had everyone stand back, closed the main enclosure door, and turned the main disconnect on.

Gordon then sequentially engaged all the branch fusing and breakers, starting from the highest values closest to the disconnect. Since he had checked all the circuits himself using his meter, he knew there should be no problems, but it was always safest to follow good practices.

All the E-Stops were pulled to their disengaged positions, and Gordon pushed the Power On/Reset button. The button illuminated, the MCR engaged, and the machine was finished with its power-up procedure.

9.6.2 Packaging Integration

Before the program was downloaded, all the auxiliary equipment was powered on and tested. The conveyors were driven by VFDs, so the keypad on the front was used to turn the conveyors on. This ensured that the motors would turn in the correct direction; if they did not it was easy to go into the parameter settings on the drives. If motor starters had been used, the phases to the starters would have had to be swapped. Speed control of the conveyors was needed, though, so VFDs had been chosen.

The vibratory bowl was quite simple and started right up. The filler would not seem to run no matter what was done, so Gordon arranged for a technician to come out the next day.

When the technician arrived, he explained that the E-Stop circuit had to be wired into the main line and the jumpers for auxiliary equipment had been inadvertently left out. After the jumpers were in place and the E-Stop circuit wired into the line, the technician ran the filler through its paces.

After all the stand-alone elements had been set up and exercised, it was time to begin integrating the entire line.

9.6.3 Controls

The standard equipment layout for LocalTech included a computer utility port on the front of the enclosure. This was manufactured by Grace Engineered Products and was somewhat generically called a "Graceport." This was a standard duplex electrical utility outlet with a GFCI; it also included a communications port that could be ordered in various configurations. This one had an RJ45 Ethernet port connected to an internal Ethernet switch linking the PLC, HMI, and Ethernet/IP I/O devices.

Gordon first used a BootP utility to set the addresses on the PLC and HMI. After communications was established, he downloaded both the PLC and HMI programs.

As usual there were a few indicators on the HMI that did not connect to the right points in the PLC. These were easy to spot since they showed up as dark blue spots on the screen. Gordon corrected the addresses and went through all the screens to ensure that everything flowed properly.

Usually Gordon used an emulator to test the PLC and HMI programs during the writing of the software, but there had simply not been enough time. Since this was a smaller line, it was really just as easy to check the software on the machine anyway.

Gordon brought up the I/O screens and showed the electrician how to navigate the HMI. He gave the electrician an I/O list and had him go to every sensor and actuator and ensure they worked all the way through the system.

Since the E-Stop circuit had been checked and actuator movements adjusted, Gordon then placed the line in manual mode. After checking the light stack and ensuring the yellow light was illuminated, he went through the process of pressing all the actuator buttons and ensuring movement. He started and stopped the chassis drive without the clutch engaged and then went ahead and restarted to observe the chassis operation. So far everything had worked without a hitch!

As a standard part of the PLC template, LocalTech had a mode called "dry cycle." This allowed the line to be operated without any product loaded just to exercise all the actuators. The next morning Gordon decided to let the machine dry cycle for a few hours just to make sure that nothing was loose. After lunch the electrician told Gordon that the machine had run pretty continuously except for a prox that had needed to be adjusted and a loose bolt on a pusher. That afternoon it was time to load a few widgets into the machine and see how the machine reacted.

It took nearly two weeks of work before Gordon and Joe were satisfied that the machine was ready for full automatic operation. The vibratory bowl was loaded with widgets provided by ACME, and they ran the machine for several more days.

Mark, the project engineer from ACME, had been talking with Gordon and Joe regularly since the start of the project. He had stopped by during the assembly of the line and stopped in again to watch the operation of the line in auto. Everyone agreed that it was time to talk about running the factory acceptance test (FAT).

9.7   FAT and SAT

9.7.1   Factory Acceptance

Mark and the production manager, along with two experienced line operators from line B, showed up bright and early Monday morning. The FAT for this line was to be a continuous four hour run with at least 95 percent uptime and full specified speed of 18 parts per hour.

A major element of the speed requirement was that the operators had to perform their tasks in concert with the machine. Exceptions for bathroom breaks and machine maintenance were a part of the overall plan, but for this run the line operation needed to be continuous. Rejected parts would also be checked for conformance and known bad components inserted by the operators.

Check sheets with a list of performance criteria had been created by ACME with input from LocalTech for both this runoff and the subsequent one after installation.

After a nearly flawless run the maintenance technicians and operators ran the machine through some manual and calibration procedures as part of their training process. Mark had no qualms in signing off the line as being ready for shipment.

9.7.2   Site Acceptance

After the machine had been shipped and installed, the site acceptance test (SAT) was performed similarly to the FAT. An eight-hour run (a full shift) was performed using additional operators from line B. The line seemed to run even better with over 98 percent uptime.

During the initial start-up at ACME, there had been some issues with the Cognex vision system. Gordon had found out that light from the Mercury Vapor overhead lighting was casting a reflected beam into the inspection area; a quickly fabricated shield was placed over the guarding and documented for inclusion in the maintenance manual.

9.8 Installation

9.8.1 Shipping

After the FAT, the line was taken apart into its individual machines. The original crating for the vibratory bowl and packaging machines had been set aside and were brought back out for reshipping.

The main chassis was unbolted from the floor and placed on wooden "4 × 4" skids. The machine was picked up one end at a time using a forklift to slide the skidding underneath, and the legs were bolted to the skidding.

Since the machine only had to be shipped across town, it was loaded onto a local truck without any crating. Straps were used to tie it off, and the other machines and various boxes of odds and ends were loaded on the same truck.

9.8.2 Contract Millwright and Electrician

LocalTech had formed a relationship with a local mechanical millwright and industrial electrician for installations. They met the truck at the ACME plant along with Gordon and Joe and assisted with getting the crates and skids to the installation area.

Unlike the assembly area at LocalTech, production areas did not always have overhead bridge cranes or forklifts to lift heavy equipment. The millwright had assisted with the installation of many of LocalTech's machines in the past and could be counted on to treat the equipment carefully.

After getting the equipment into the right location on the plant floor, the crating and skidding were removed. The machines were all releveled and bolted to the floor using a hammer drill and concrete anchors. Special concrete pads had been poured to drill into for part of the framing.

With the help of LocalTech's electrician and the industrial electrical contractor, the machines were reconnected. New conduit was run from the main panel to the individual machines for power; this had been temporary at the LocalTech assembly floor.

ACME's facility engineer came out and ensured that power and air were dropped to the proper location for the line. The machines were powered up much more quickly than during the original start-up since the system had already been checked thoroughly. After

Gordon and Joe did a quick check of the machinery, they powered up the line and prepared for SAT.

9.9 Support

After the SAT two weeks went by while ACME performed training and qualification of the equipment. Mark spent nearly all his time out on the floor as two shifts had to be brought up to speed. After two weeks the line went into full production as line C.

9.9.1 The First Three Months

There was no doubt that line C was faster and more high-tech than line B. Despite the fact that LineX had more experience with the widget manufacturing industry, LocalTech had proven to be very technical and detail oriented.

The OEE screens on the HMI were helpful in determining the causes of stoppages in specific areas of the line. The quality manager in particular mentioned that he would like to see this implemented on lines A and B also if the budget would allow for it.

Gordon and Joe had needed to come by several times during the first few months to make some minor adjustments and software changes, but overall the project had turned out to be a huge success. LocalTech would be seeing more business from ACME on future projects for sure!

9.9.2 Warranty

About nine months into the operation of the line, one of the AC motors on a conveyor had stopped working. LocalTech was called and the motor sent to the manufacturer for replacement. Since the motor had been on the recommended spare parts list, it was quickly replaced and production resumed.

One of the photo-eyes on the outfeed conveyor had also mysteriously been found broken in the morning twice (after night shift, of course). Mark had a guard fabricated and bolted over the sensor and anticipated no further issues.

CHAPTER 10

Applications

These are some of the machines and systems the author has designed or been involved with over the past 20 years or so.

10.1 Binder-Processing Machine

Customer: Avery Dennison
Completed: 1999
Mechanical: Nalle Automation Systems (NAS)
Controls: Automation Consulting (ACS)
Frame: Aluminum Extrusion
Controller: Allen-Bradley SLC 5/04
Other Components: Emerson Servo

This is a machine built by NAS in the 1998 to 1999 time frame. As you can tell, NAS was a huge user of aluminum extrusion; for about five years or so they were the biggest 80–20 customer in East Tennessee. This machine was built for Avery Dennison in Chicopee, Massachusetts. It took the components of a three-ring binder—rings, inserts, labels, and so on—assembled them, arranged them in alternating stacks, and put them in a box for shipping.

At least five of these machines were built over about a three-year period. Probably the most interesting feature of the machine was the method of indexing binders from station to station. The binders slid into a loading station; then a servo actuated mechanism slid a pair of spring-loaded fingers behind the back edge of the binder and pushed it from station to station. As the indexer returned from its push, the fingers were pressed down beneath the binder surface by the binder itself; as the finger emerged behind the binder, the spring would pop it up again.

Whenever relying on a purely mechanical method such as this plenty of sensors and monitoring of torque needs to be used to detect the inevitable crashes. The product is not always completely uniform and operator loading error could also be a factor. Quite a few binders were destroyed in the runoff process. If they were not in too bad of shape, they were used as project notebooks.

Chapter Ten

FIGURE 10.1 Binder-processing machine.

The drive system used for this system was an Emerson Servo. The nice thing about this particular servo was that it was quite self-contained and had a very user-friendly software and configuration package. Integrated servo packages existed within the A-B SLC500 platform that we were using, but they were much less user-friendly than they are with the Kinetix and Sercos systems used today. A-B also had the Ultra stand-alone systems at that time, but they were still much less user-friendly.

At a basic level this machine was still a packaging machine despite the assembly parts of it. NAS eventually moved on to building primarily standard packaging equipment.

10.2 Crystal Measurement

Customer: CTI
Completed: 2005
Mechanical: Agile Engineering
Controls: Automation Consulting (ACS)
Frame: Welded Steel Base, Aluminum Extrusion Guarding
Controller: Koyo DL205
Other Components: Keyence Laser Scanner

Applications 307

FIGURE 10.2 Crystal gauging station.

This station shown in Fig. 10.2 was used to measure a small crystal used in the manufacture of medical CT scanners. A Keyence laser scanner was used to measure the crystal in three different axes, providing both dimensional and "squareness" information to an operator.

The controller used was a Koyo (Automation Direct) PLC and a Keyence precision scanning CCD laser measurement system. The operator unit on the right shows the Keyence unit where set points and measurement data are set up. This information is transferred serially to the PLC, which logs the data into registers for transfer to a computer with a database. This technique is often used in multiple measurement applications where a PLC is used as a manipulator controller and a data concentrator.

The crystal (shown in Fig. 10.3 to the left of the stainless steel datum—it is very small) is placed in the tooling, which contains a small vacuum hole. The vacuum pulls the crystal up firmly into a small pocket. An analog vacuum measurement is used to determine whether the top of the crystal is flat; irregularities cause the vacuum to leak. The steel datum point is used as a reference to measure the length of the crystal precisely rather than using the actual position of the actuators.

Figure 10.3 Crystal manipulator and measurement.

This provides accuracy greater than if the actual position were used because of the slop in the actuator bearing. The distance from the side of the crystal to the datum is also measured at the top and bottom of the crystal to give the parallelness of the top to the sides. The crystal is rotated 90° on axis and the measurement taken again. The entire mechanism then tilts 90° to measure the length.

There is a lot going on in a very small space in this application. The tooling was designed by Agile Engineering in Knoxville, Tennessee, for the end user, CTI. My old company, ACS, designed and built the controls. Another example of several companies working together for a successful project.

10.3 SmartBench

Customer: None (Prototype)

Completed: 2003

Mechanical and Controls: Automation Consulting (ACS)

Frame: Aluminum Extrusion
Controller: Koyo DL205
Other Components: Maple Systems Touchscreen

The SmartBench is a combination of software and hardware controls that allows the programming of sequential events by means of entering functions by number into step locations on a touch screen. Functions may react to specified inputs, energize or deenergize outputs, perform timed "waits," or detect timed or out of sequence faults. As configured in Fig. 10.4, there are 32 digital inputs and 32 digital outputs grouped into eight 4in/4out cable ports, which can be connected to configured stations such as the assembly pallet shown above. Functions are then allowed to be named by the programmer by means of a pop-up alphanumeric touch screen on the HMI.

The system will automatically start at the first programmed step or function, proceeding to the next when the current step's completion is accomplished. This will continue until a nonprogrammed or zero step is encountered, after which the sequence will start over.

FIGURE 10.4   The SmartBench.

The operating system of the SmartBench shown above is a Koyo/Automation Direct PLC with two 16-point input and two 16-point output cards. There is also a slot-mounted Ethernet card installed for remote download of recipes or sequences from a PC, allowing the sequence to be arranged by means of a spreadsheet or database program such as Microsoft Excel or Access.

The current onboard operator interface or HMI is a Maple Systems touch screen with onboard recipe memory. The PLC performs the logic, sequencing, and timing functions and rearranging of function names into their proper step locations, fault detection, and password access to screens. The operator interface stores the configured sequences and allows entry of alphanumeric data descriptions into the PLC registers associated with each function. The picture in Fig. 10.4 was taken at an industrial trade show. The HMI touchscreen is to the right; the computer on the left was used for the attached vision system.

The SmartBench was created as a replacement for several dedicated purpose assembly stations at TRW Koyo, Vonore, Tennessee. It was meant to be able to individually replace any of the existing machines as they were retooled for the next program year of hose assemblies. At the Greenville tradeshow, a pallet was built where attendees could assemble a balsa airplane. As the first part, the body, was inserted into the tooling, a latch was activated, locking the piece in place. Lights would blink on both the bins and the tooling indicating where the operator was to pick or place the next component. After all parts were detected to be in place, a camera was triggered to inspect the plane for items such as upside-down wings or missing components. After successful completion, the plane would be released and the sequence started over. The trade show was quite successful in acquiring new projects, although most were related to other applications.

Although a Koyo PLC and a Maple Systems touch screen were used for the prototype SmartBench, the programming techniques can be used on any platform that has the same capabilities. The most cost-effective form would be to put the operating system on a microprocessor and create a circuit card that could interface with I/O.

10.4 Sagger Load Station

Customer: Alcoa/Howmet

Completed: 2005

Mechanical and Controls: Automation Consulting (ACS)

Frame: Aluminum Extrusion

Controller: Allen Bradley SLC5/05

Other Components: Maple Systems Touchscreen, Motoman HP6 Robot

Applications

FIGURE 10.5 Sagger Load Station.

The Sagger Load Station is a system that loads ceramic turbine blade castings into a ceramic box filled with sand called a "sagger." Loaded saggers are placed on an infeed conveyor, where they are indexed into a pick-and-place station that inverts the sagger over a screen, allowing the sand to fall into a hopper. The empty sagger is then indexed around to the backside of the hopper, where a layer of sand is fed by a screw feeder from the bottom of the hopper into the sagger. The sagger indexes to a load position, where a robot removes ceramic castings from a mold and gently embeds it into the sand layer. Once the layer of castings is complete, the sagger is indexed back to the hopper station and another layer of sand is poured over the casting layer. After indexing back under the robot, a layer of castings is inserted, then a layer of sand until the sagger is loaded. A final layer of sand is placed on the top, and the sagger exits the station. Operators then place the saggers in an oven for several hours; the sagger is brought back to the station and unloaded and the sand-filled sagger placed on the infeed conveyor again.

A unique feature of this station is the spring-loaded cammed pick-and-place inverter. As the saggers were quite heavy, a mechanism was used to balance the sagger load at every point as it was inverted, allowing a much smaller rotary pneumatic cylinder to be used for both forward and reverse motions.

This system was installed in early 2005 and was one of ACS's first large machines. It used an Allen-Bradley SLC5/05 processor and a Maple Systems touch screen. A MotoMan HP6 Robot was used to unload the molding press and place castings into the sagger.

312 Chapter Ten

FIGURE 10.6 Tray handler.

10.5 Tray Handlers

Customer: Undisclosed
Completed: 2009
Mechanical and Controls: Wright Industries
Frame: Welded Steel
Controller: Siemens S7
Other Components: WinCC Touch Screen, Siemens Sinamics Servos

Figure 10.6 is an illustration of a nonrobotic motion control application. As with most of my other motion jobs, this one did not have any kinds of coordinated motion involved, but it was challenging from a sensing standpoint. The concept was to be able to stack (fill) or destack (empty) trays. An operator would load the tray handler with an either full or empty stack of trays with a forklift. The trays would be presented one at a time to an operator, starting either from the bottom or the top of the stack. Much of the coding involved teaching or programming the various stack heights, tray positions, and lift points and remembering where you were in the process regardless of what manual activities had taken place.

Two parallel pairs of Z-axes were used to lift the trays. One was the tray lift, which would move to the tray above that which was to be removed. It would extend fingers into slots in the trays and then lift the stack off the target tray. The tray extender would then extend its fingers, lift slightly, and a horizontal axis would extend the tray forward to be either filled or emptied by the operator. The two Z-axes and the horizontal axis were all Siemens Sinamics servo drives and motors controlled by a Siemens S7-300 PLC. All drive and HMI communications were via Profibus.

The trickiest part of this application was in knowing where the trays were. Since the tray stacks were always full of trays of the same thickness, all the tray positions were simply programmed as an index value. This saved money from the standpoint of not having to put sensors on the fingers or scan the stack, but it caused some problems when stacks that did not conform were inserted. The trays and spacer poles were made of graphite, a brittle ceramic material that could withstand high oven temperatures; as such they were quite breakable. A few photo-eyes were located in the machine for product location, but they did not always work where trays were misoriented.

Overtravel switches were located in the usual locations near the ends of the axes, but ideally they would have been put on the vertical axes tooling mounts themselves since the axes could hit each other long before they reached end of stroke. This is a good thing to remember whenever working with two parallel axes that can interfere with each other: always take into account what else the axis can hit besides end of travel on the actuator itself.

10.6 Cotton Classing System

Customer: U.S. Department of Agriculture (USDA)

Completed: 2000

Controls: Automation Consulting (ACS)

Conveyors: Nalle Automation (NAS)

Testers: Zellweger-Uster

Frame: Aluminum Extrusion

Controller: Allen-Bradley SLC5/04

Other Components: Rockwell RSView32 on computer, embedded controllers on testers

314   Chapter Ten

Figure 10.7   Cotton classing system.

The picture in Fig. 10.7 is of the cotton classing system at the USDA facility in Memphis. This was a cooperative project between my company ACS, NAS, and Zellweger-Uster. This is an example of automation that is not located in your typical industrial facility.

Often companies or organizations without a great deal of automation history reach a point where they feel they need to automate some of their processes to save money or increase production. The people who are part of the processes being automated are usually highly skilled at their tasks but do not have experience with automation techniques and concepts. Devices like HMIs, E-Stops, and light curtains may be new to the facility, and there is often no maintenance staff skilled in the troubleshooting and care of these systems. There is also the incongruity of noisy industrial equipment being placed into a lab or clean room–type environment.

In this case, the product being handled and inspected is cotton samples from growers throughout the Southeast. The picture back in Fig. 4.1, shows a cotton sample being transported on this system. Prior to using this system, small pieces of cotton samples were placed under cameras by hand to determine color and the quantity of foreign matter and seeds in the sample. This system allowed the material handling to be done without human intervention. As with many other projects, there were a lot of iterations of the system, and it grew more complex as lessons were learned. A third level of conveyors was added to allow retesting of the samples upon inconclusive results.

Because of the loose tufts of cotton that inevitably ended up floating around the facility, there was talk of entirely enclosing the cotton sample pallets; I believe the system was decommissioned before this was ever done.

Because of this same incongruity of having semi-industrial equipment effectively installed on a tile floor in the equivalent of an office building, the system eventually was removed and operators went back to loading test samples by hand. There are some things that just cannot be automated effectively, was this one of them? Only time will tell....

APPENDIX A
ASCII Table

ASCII stands for American Standard Code for Information Interchange. Computers can only understand numbers, so an ASCII code is the numerical representation of a character such as *a* or @ or an action of some sort. ASCII was developed a long time ago and now the nonprinting characters are rarely used for their original purpose. Below is the ASCII character table, which includes descriptions of the first 32 nonprinting characters. ASCII was actually designed for use with teletypes, so the descriptions are somewhat obscure. If someone says they want your document in ASCII format, all this means is they want "plain" text with no formatting such as tabs, bold, or underscoring—the raw format that any computer can understand. This is usually so they can easily import the file into their own applications without issues. Notepad.exe creates ASCII text, or in MS Word you can save a file as "text only."

Dec	Hex	Character	Name
0	0	NUL	Null
1	1	STX	Start of Header
2	2	SOT	Start of Text
3	3	ETX	End of Text
4	4	EOT	End of Transmission
5	5	ENQ	Enquiry
6	6	ACK	Acknowledge
7	7	BEL	Bell
8	8	BS	BackSpace
9	9	HT	Horizontal Tabulation
10	0A	LF	Line Feed
11	0B	VT	Vertical Tabulation

TABLE A1 Basic ASCII

Dec	Hex	Character	Name
12	0C	FF	Form Feed
13	0D	CR	Carriage Return
14	0E	SO	Shift Out
15	0F	SI	Shift In
16	10	DLE	Data Link Escape
17	11	DC1	Device Control 1 (XON)
18	12	DC2	Device Control 2
19	13	DC3	Device Control 3 (XOFF)
20	14	DC4	Device Control 4
21	15	NAK	Negative acknowledge
22	16	SYN	Synchronous Idle
23	17	ETB	End of Transmission Block
24	18	CAN	Cancel
25	19	EM	End of Medium
26	1A	SUB	Substitute
27	1B	ESC	Escape
28	1C	FS	File Separator
29	1D	GS	Group Separator
30	1E	RS	Record Separator
31	1F	US	Unit Separator
32	20	[Space]	Space
33	21	!	Exclamation mark
34	22	"	Quotes
35	23	#	Hash
36	24	$	Dollar
37	25	%	Percent
38	26	&	Ampersand
39	27	'	Apostrophe
40	28	(	Open bracket
41	29	)	Close bracket
42	2A	*	Asterisk
43	2B	+	Plus
44	2C	,	Comma

TABLE A1 Basic ASCII (*Continued*)

Dec	Hex	Character	Name
45	2D	-	Dash
46	2E	.	Full stop
47	2F	/	Slash
48	30	0	Zero
49	31	1	One
50	32	2	Two
51	33	3	Three
52	34	4	Four
53	35	5	Five
54	36	6	Six
55	37	7	Seven
56	38	8	Eight
57	39	9	Nine
58	3A	:	Colon
59	3B	;	Semi-colon
60	3C	<	Less than
61	3D	=	Equals
62	3E	>	Greater than
63	3F	?	Question mark
64	40	@	At
65	41	A	Uppercase A
66	42	B	Uppercase B
67	43	C	Uppercase C
68	44	D	Uppercase D
69	45	E	Uppercase E
70	46	F	Uppercase F
71	47	G	Uppercase G
72	48	H	Uppercase H
73	49	I	Uppercase I
74	4A	J	Uppercase J
75	4B	K	Uppercase K
76	4C	L	Uppercase L
77	4D	M	Uppercase M

TABLE A1 Basic ASCII (*Continued*)

Appendix A

Dec	Hex	Character	Name
78	4E	N	Uppercase N
79	4F	O	Uppercase O
80	50	P	Uppercase P
81	51	Q	Uppercase Q
82	52	R	Uppercase R
83	53	S	Uppercase S
84	54	T	Uppercase T
85	55	U	Uppercase U
86	56	V	Uppercase V
87	57	W	Uppercase W
88	58	X	Uppercase X
89	59	Y	Uppercase Y
90	5A	Z	Uppercase Z
91	5B	[	Open square bracket
92	5C	\	Backslash
93	5D	]	Close square bracket
94	5E	^	Caret / hat
95	5F	_	Underscore
96	60	`	Grave accent
97	61	a	Lowercase a
98	62	b	Lowercase b
99	63	c	Lowercase c
100	64	d	Lowercase d
101	65	e	Lowercase e
102	66	f	Lowercase f
103	67	g	Lowercase g
104	68	h	Lowercase h
105	69	i	Lowercase i
106	6A	j	Lowercase j
107	6B	k	Lowercase k
108	6C	l	Lowercase l
109	6D	m	Lowercase m
110	6E	n	Lowercase n

TABLE A1 Basic ASCII (*Continued*)

ASCII Table

Dec	Hex	Character	Name
111	6F	o	Lowercase o
112	70	p	Lowercase p
113	71	q	Lowercase q
114	72	r	Lowercase r
115	73	s	Lowercase s
116	74	t	Lowercase t
117	75	u	Lowercase u
118	76	v	Lowercase v
119	77	w	Lowercase w
120	78	x	Lowercase x
121	79	y	Lowercase y
122	7A	z	Lowercase z
123	7B	{	Open brace
124	7C	|	Pipe
125	7D	}	Close brace
126	7E	~	Tilde
127	7F	DEL	Delete

TABLE A1 Basic ASCII (Continued)

Dec	Hex	Character	Description
128	80	Ç	Latin capital letter C with cedilla
129	81	ü	Latin small letter u with diaeresis
130	82	é	Latin small letter e with acute
131	83	â	Latin small letter a with circumflex
132	84	ä	Latin small letter a with diaeresis
133	85	à	Latin small letter a with grave
134	86	å	Latin small letter a with ring above
135	87	ç	Latin small letter c with cedilla
136	88	ê	Latin small letter e with circumflex
137	89	ë	Latin small letter e with diaeresis
138	8A	è	Latin small letter e with grave

TABLE A2 Extended ASCII

Appendix A

Dec	Hex	Character	Description
139	8B	ï	Latin small letter i with diaeresis
140	8C	î	Latin small letter i with circumflex
141	8D	ì	Latin small letter i with grave
142	8E	Ä	Latin capital letter A with diaeresis
143	8F	Å	Latin capital letter A with ring above
144	90	É	Latin capital letter E with acute
145	91	æ	Latin small ligature ae
146	92	Æ	Latin capital ligature ae
147	93	ô	Latin small letter o with circumflex
148	94	ö	Latin small letter o with diaeresis
149	95	ò	Latin small letter o with grave
150	96	û	Latin small letter u with circumflex
151	97	ù	Latin small letter u with grave
152	98	ÿ	Latin small letter y with diaeresis
153	99	Ö	Latin capital letter O with diaeresis
154	9A	Ü	Latin capital letter U with diaeresis
155	9B	¢	cent sign
156	9C	£	pound sign
157	9D	¥	yen sign
158	9E	₧	peseta sign
159	9F	ƒ	Latin small letter f with hook
160	A0	á	Latin small letter a with acute
161	A1	í	Latin small letter i with acute
162	A2	ó	Latin small letter o with acute
163	A3	ú	Latin small letter u with acute
164	A4	ñ	Latin small letter n with tilde
165	A5	Ñ	Latin capital letter n with tilde
166	A6	ª	feminine ordinal indicator
167	A7	º	masculine ordinal indicator
168	A8	¿	inverted question mark
169	A9	⌐	reversed not sign
170	AA	¬	not sign
171	AB	½	vulgar fraction one half

TABLE A2 Extended ASCII (*Continued*)

ASCII Table

Dec	Hex	Character	Description
172	AC	¼	vulgar fraction one quarter
173	AD	¡	inverted exclamation mark
174	AE	«	left-pointing double angle quotation mark
175	AF	»	right-pointing double angle quotation mark
176	B0		light shade
177	B1		medium shade
178	B2		dark shade
179	B3	│	box drawings light vertical
180	B4	┤	box drawings light vertical and left
181	B5	╡	box drawings vertical single and left double
182	B6	╢	box drawings vertical double and left single
183	B7	╖	box drawings down double and left single
184	B8	╕	box drawings down single and left double
185	B9	╣	box drawings double vertical and left
186	BA	║	box drawings double vertical
187	BB	╗	box drawings double down and left
188	BC	╝	box drawings double up and left
189	BD	╜	box drawings up double and left single
190	BE	╛	box drawings up single and left double
191	BF	┐	box drawings light down and left
192	C0	└	box drawings light up and right
193	C1	┴	box drawings light up and horizontal
194	C2	┬	box drawings light down and horizontal
195	C3	├	box drawings light vertical and right
196	C4	─	box drawings light horizontal
197	C5	┼	box drawings light vertical and horizontal
198	C6	╞	box drawings vertical single and right double

TABLE A2 Extended ASCII (*Continued*)

Appendix A

Dec	Hex	Character	Description
199	C7	╟	box drawings vertical double and right single
200	C8	╚	box drawings double up and right
201	C9	╔	box drawings double down and right
202	CA	╩	box drawings double up and horizontal
203	CB	╦	box drawings double down and horizontal
204	CC	╠	box drawings double vertical and right
205	CD	═	box drawings double horizontal
206	CE	╬	box drawings double vertical and horizontal
207	CF	╧	box drawings up single and horizontal double
208	D0	╨	box drawings up double and horizontal single
209	D1	╤	box drawings down single and horizontal double
210	D2	╥	box drawings down double and horizontal single
211	D3	╙	box drawings up double and right single
212	D4	╘	box drawings up single and right double
213	D5	╒	box drawings down single and right double
214	D6	╓	box drawings down double and right single
215	D7	╫	box drawings vertical double and horizontal single
216	D8	╪	box drawings vertical single and horizontal double
217	D9	┘	box drawings light up and left
218	DA	┌	box drawings light down and right
219	DB	█	full block
220	DC	▄	lower half block
221	DD	▌	left half block
222	DE	▐	right half block

TABLE A2 Extended ASCII (*Continued*)

ASCII Table

Dec	Hex	Character	Description
223	DF	■	upper half block
224	E0	α	Greek small letter alpha
225	E1	ß	latin small letter sharp s
226	E2	Γ	Greek capital letter gamma
227	E3	π	Greek small letter pi
228	E4	Σ	Greek capital letter sigma
229	E5	σ	Greek small letter sigma
230	E6	µ	micro sign
231	E7	τ	Greek small letter tau
232	E8	Φ	Greek capital letter phi
233	E9	Θ	Greek capital letter theta
234	EA	Ω	Greek capital letter omega
235	EB	δ	Greek small letter delta
236	EC	∞	infinity
237	ED	φ	Greek small letter phi
238	EE	ε	Greek small letter epsilon
239	EF	∩	intersection
240	F0	≡	identical to
241	F1	±	plus-minus sign
242	F2	≥	greater-than or equal to
243	F3	≤	less-than or equal to
244	F4	⌠	top half integral
245	F5	⌡	bottom half integral
246	F6	÷	division sign
247	F7	≈	almost equal to
248	F8	°	degree sign
249	F9	·	bullet operator
250	FA	·	middle dot
251	FB	√	square root
252	FC	ⁿ	superscript latin small letter n
253	FD	²	superscript two
254	FE	■	black square
255	FF		no-break space

TABLE A2 Extended ASCII (*Continued*)

APPENDIX B
Ampacity

Ampacity data sourced from Square D Motor Data Calculator based on the National Electrical Code (NEC).

Copper Wire Ampacity	
AWG	Ampacity
14	20
12	25
10	35
8	50
6	65
4	85
3	100
2	115
1	130
1/0	150
2/0	175
3/0	200
4/0	230
250	255
300	285
350	310
400	335
500	380
600	420
700	460
750	475
800	490
900	520
1000	545

TABLE B1 Wire Ampacity

APPENDIX C
Motor Sizing

Motor data sourced from Square D Motor Data Calculator based on the National Electrical Code (NEC).

Three-Phase Motor Data
For 60 Hz 1800RPM Standard Squirrel Cage Motors (Non Design E)

HP	200(208) Volts				230(240) Volts			
	FLA	Min. Copper Wire Size	Circuit Breaker	Fusible Switch	FLA	Min. Copper Wire Size	Circuit Breaker	Fusible Switch
0.50	2.2	14	15	4	2.2	14	15	4
0.75	3.7	14	15	6.25	3.2	14	15	5.6
1.00	4.8	14	15	8	4.2	14	15	8
1.50	6.9	14	15	10	6.0	14	15	10
2.00	7.8	14	15	10	6.8	14	15	10
3.00	11.0	14	20	17.5	9.6	14	20	15
5.00	17.5	12	35	25	15.2	14	30	25
7.50	25.3	10	50	40	22.0	10	45	30
10.00	32.2	8	60	50	28.0	10	60	40
15.00	48.3	6	90	60	42.0	6	80	60
20.00	62.1	4	100	90	54.0	4	90	80
25.00	78.2	3	110	100	68.0	4	100	100
30.00	92.0	2	125	125	80.0	3	110	100
40.00	120.0	1/0	175	175	104.0	1	150	150
50.00	150.0	3/0	200	200	130.0	2/0	200	200

TABLE C1 Three-Phase Motor Sizing

HP	FLA	Min. Copper Wire Size	Circuit Breaker	Fusible Switch	FLA	Min. Copper Wire Size	Circuit Breaker	Fusible Switch
60.00	177.0	4/0	250	250	154.0	3/0	225	200
75.00	221.0	300	300	300	192.0	250	250	300
100.00	285.0	500	400	400	248.0	350	350	350
125.00	359.0	2-4/0	600	500	312.0	2-3/0	450	400
150.00	414.0	2-300	600	600	360.0	2-4/0	600	500
200.00	552.0	2-500	800	N/A	480.0	2-500	800	600

Three-Phase Motor Data
For 60 Hz 1800RPM Standard Squirrel Cage Motors (Non Design E)

HP	460(480) Volts				575(600) Volts			
	FLA	Min. Copper Wire Size	Circuit Breaker	Fusible Switch	FLA	Min. Copper Wire Size	Circuit Breaker	Fusible Switch
0.50	1.1	14	15	2	0.9	14	15	1.8
0.75	1.6	14	15	3.2	1.3	14	15	2.5
1.00	2.1	14	15	4	1.7	14	15	3.2
1.50	3.0	14	15	5.6	2.4	14	15	4
2.00	3.4	14	15	6.25	2.7	14	15	5
3.00	4.8	14	15	8	3.9	14	15	6.25
5.00	7.6	14	15	15	6.1	14	15	10
7.50	11.0	14	20	20	9.0	14	15	15

TABLE C1 Three-Phase Motor Sizing (*Continued*)

Three-Phase Motor Data
For 60 Hz 1800RPM Standard Squirrel Cage Motors (Non Design E)

HP	460(480) Volts				575(600) Volts			
	FLA	Min. Copper Wire Size	Circuit Breaker	Fusible Switch	FLA	Min. Copper Wire Size	Circuit Breaker	Fusible Switch
10.00	14.0	14	25	20	11.0	14	20	20
15.00	21.0	10	40	30	17.0	12	35	25
20.00	27.0	10	60	40	22.0	10	45	30
25.00	34.0	8	70	50	27.0	10	60	40
30.00	40.0	8	80	60	32.0	8	60	50
40.00	52.0	6	90	80	41.0	6	80	60
50.00	65.0	4	100	100	52.0	6	90	80
60.00	77.0	3	110	100	62.0	4	100	90
75.00	96.0	1	125	150	77.0	3	110	100
100.00	124.0	2/0	200	175	99.0	1	150	150
125.00	156.0	3/0	225	200	125.0	2/0	200	175
150.00	180.0	4/0	250	250	144.0	3/0	200	200
200.00	240.0	350	350	350	192.0	250	250	300

TABLE C1 Three-Phase Motor Sizing (Continued)

Single-Phase Motor Data

For 60 Hz 1800RPM Standard Squirrel Cage Motors (Non Design E)

HP	115(120) Volts				230(240) Volts			
	FLA	Min. Copper Wire Size	Circuit Breaker	Fusible Switch	FLA	Min. Copper Wire Size	Circuit Breaker	Fusible Switch
1/6	4.4	14	15	6.25	2.2	14	15	3.2
1/4	5.8	14	15	9	2.9	14	15	4.5
1/3	7.2	14	15	10	3.6	14	15	5.6
1/2	9.8	14	20	15	4.9	14	15	7
3/4	13.8	14	25	20	6.9	14	15	10
1.00	16.0	14	30	25	8.0	14	15	12
1.50	20.0	12	40	30	10.0	14	20	15
2.00	24.0	10	50	30	12.0	14	25	20
3.00	34.0	8	70	50	17.0	12	35	25
5.00	56.0	4	90	80	28.0	10	60	40
7.50	80.0	3	110	100	40.0	8	80	60
10.00	*	*	*	*	50.0	6	90	60

TABLE C2 Single-Phase Motor Sizing

APPENDIX D
NEMA Enclosure Tables

Comparison of Specific Applications of Enclosures for Indoor Nonhazardous Locations
[Table 1 from NEMA 250-2003]

Provides a Degree of Protection against the Following Conditions	Type of Enclosure									
	1*	2*	4	4X	5	6	6P	12	12K	13
Access to hazardous parts	X	X	X	X	X	X	X	X	X	X
Ingress of solid foreign objects (Falling dirt)	X	X	X	X	X	X	X	X	X	X
Ingress of water (Dripping and light splashing)	...	X	X	X	X	X	X	X	X	X
Ingress of solid foreign objects (Circulating dust, lint, fibers, and filings **)	...	...	X	X	...	X	X	X	X	X
Ingress of solid foreign objects (Settling airborne dust, lint, fibers, and flyings **)	...	...	X	X	X	X	X	X	X	X
Ingress of water (Hosedown and splashing water)	...	...	X	X	...	X	X	...	...	...
Oil and coolant seepage	...	...	...	...	...	...	...	X	X	X
Oil or coolant spraying and splashing	...	...	...	...	...	...	...	...	...	X
Corrosive agents	...	...	...	X	...	...	X	...	...	...
Ingress of water (Occasional temporary submersion)	...	...	...	...	...	X	X	...	...	...
Ingress of water (Occasional prolonged submersion)	...	...	...	...	...	...	X	...	...	...

* These enclosures may be ventilated.
** These fibers and flyings are nonhazardous materials and are not considered Class III type ignitable fibers or combustible flyings. For Class III type ignitable fibers or combustible flyings see the National Electrical Code, Article 500.

TABLE D1 NEMA Enclosure Ratings—Indoor Nonhazardous

Comparison of Specific Applications of Enclosures for Outdoor Nonhazardous Locations
[Table 2 from NEMA 250-2003]

Provides a Degree of Protection against the Following Conditions	Type of Enclosure									
	3	3X	3R*	3RX*	3S	3SX	4	4X	6	6P
Access to hazardous parts	X	X	X	X	X	X	X	X	X	X
Ingress of water (Rain, snow, and sleet **)	X	X	X	X	X	X	X	X	X	X
Sleet ***	...	...	...	...	X	X	...	...	...	...
Ingress of solid foreign objects (Windblown dust, lint, fibers, and flyings)	X	X	...	...	X	X	X	X	X	X
Ingress of water (Hosedown)	...	...	...	...	...	...	X	X	X	X
Corrosive agents	...	X	...	X	...	X	...	X	...	X
Ingress of water (Occasional temporary submersion)	...	...	...	...	...	...	...	...	X	X
Ingress of water (Occasional prolonged submersion)	...	...	...	...	...	...	...	...	...	X

* These enclosures may be ventilated.
** External operating mechanisms are not required to be operable when the enclosure is ice covered.
*** External operating mechanisms are operable when the enclosure is ice covered.

TABLE D2 NEMA Enclosure Ratings—Outdoor Nonhazardous

Comparison of Specific Applications of Enclosures for Indoor Hazardous Locations
(If the installation is outdoors and/or additional protection is required by Table 1 and Table 2, a combination-type enclosure is required.)
[Table B1 from NEMA 250-2003]

Provides a Degree of Protection against Atmospheres Typically Containing (See NFPA 497M for Complete Listing)	Class	Enclosure Types 7 and 8, Class I Groups **				Enclosure Type 9, Class II Groups			
		A	B	C	D	E	F	G	10
Acetylene	I	X	...	...	...	...	...	...	...
Hydrogen, manufactured gas	I	...	X	...	...	...	...	...	...
Diethyl ether, ethylene, cyclopropane	I	...	...	X	...	...	...	...	...
Gasoline, hexane, butane, naphtha, propane, acetone, toluene, isoprene	I	...	...	...	X	...	...	...	...
Metal dust	II	...	...	...	...	X	...	...	...
Carbon black, coal dust, coke dust	II	...	...	...	...	...	X	...	...
Flour, starch, grain dust	II	...	...	...	...	...	...	X	...
Fibers, flyings *	III	...	...	...	...	...	...	X	...
Methane with or without coal dust	MSHA	...	...	...	...	...	...	...	X

TABLE D3   NEMA Enclosure Ratings—Indoor Hazardous

APPENDIX E

Manufacturers, Machine Builders, and Integrators

Electrical Enclosures

Hoffman	http://www.hoffmanonline.com/
Rittal	http://rittal.com/
Saginaw	http://www.saginawcontrol.com/
Weigmann	http://hubbell-wiegmann.com/

General Control Products

Ametek	http://www.ametek.com/
	(Connectors, Instruments, Motors)
Barber-Colman	http://www.barber-colman.com/
	(Temperature Control)
Danaher	http://www.danaher.com/
	(Timers, Counters, Sensors, Motion)
Idec	http://www.idec.com/
Omega	http://omega.com/
	(Temperature Control)
Pepperl+Fuchs	http://www.pepperl-fuchs.com
Phoenix Contact	http://www.phoenixcontact.com/usa_home.htm
Wago	http://www.wago.us/

Mechanical and Structural

Bearings

Dodge	http://www.dodge-pt.com/products/bearing/
SKF	http://www.skf.com/portal/skf/home

Manufacturers, Machine Builders, and Integrators

THK http://www.thk.com/
Timken http://www.timken.com/en-us/Pages/Home.aspx

Framing and Guarding

80-20 http://8020.net/
(Aluminum Extrusion)
Bosch http://www13.boschrexroth-us.com/framing_shop/
(Aluminum Extrusion)
Creform http://www.creform.com/
(Structural Pipe Assemblies)
Item http://itemamerica.com/
(Aluminum Extrusion)
Misumi http://us.misumi-ec.com/
(Mounting and Hardware)

Indexers and Devices

Camco http://www.camcoindex.com/
DE-STA-CO http://www.destaco.com/
Robohand http://www.destaco.com/robohand-equipment.html
Stelron http://www.stelron.com/

Motion Control

Servos, Drives, and Systems

Allen-Bradley http://www.ab.com/
Bosch Rexroth http://www.boschrexroth-us.com/
Copley http://www.copleycontrols.com/
Delta Tau http://www.deltatau.com/
Emerson http://www.emersonct.com/
Faulhaber http://www.faulhaber.com/
Galil http://www.galilmc.com/
Kollmorgen http://www.kollmorgen.com/en-us/products/
Siemens http://www.industry.usa.siemens.com/
Yaskawa http://www.yaskawa.com/

Motors

Baldor http://www.baldor.com/products/
GE http://www.geindustrial.com/cwc/
Oriental http://www.orientalmotor.com/
Reliance http://www.reliance.com/
Siemens http://www.industry.usa.siemens.com/

PLCs and Controls

These manufacturers often also include sensors, motor controls, motion, wire management, and industrial devices.

ABB	http://www.abb.com/controlsystems
Allen-Bradley	http://www.ab.com/
Automation Direct	http://www.automationdirect.com/
B & R	http://www.br-automation.com/en-us/
Beckhoff	http://www.beckhoff.com/
EZAutomation	http://flash.ezautomation.net/
Foxboro	http://iom.invensys.com/EN/Pages/Foxboro.aspx
GE	http://www.ge-ip.com/
Mitsubishi	http://www.meau.com/
Modicon	http://www.modicon.com/
National Instruments	http://www.ni.com/
Omron	http://www.omron.com/
Opto 22	http://www.opto22.com/
Siemens	http://www.siemens.com/entry/cc/en/

Pneumatics and Hydraulics

Actuators, Cylinders, and Valves

Bosch Rexroth	http://www.boschrexroth.com/
Eaton	http://www.eaton.com/Eaton/index.htm
Festo	http://www.festo.com/net/StartPage/
MAC	http://www.macvalves.com/
Numatics	http://www.numatics.com/
Parker	http://www.parker.com/
SMC	http://www.smcusa.com/

Robots

ABB	http://www.abb.com/controlsystems
Adept	http://www.adept.com/
Denso	http://www.densorobotics.com/
Fanuc	http://www.fanucrobotics.com/
Kuka	http://www.kuka.com/
Motoman	http://www.motoman.com/
Panasonic	http://www.panasonicfa.com/

Sensors

Photoelectrics, Proximity, and Limit Switches

Balluff	http://www.balluff.com/Balluff/
Banner	http://www.bannerengineering.com/en-US/

Baumer http://www.baumer.com/
IFM-Efector http://www.ifmefector.com/
Keyence http://www.keyence.com/
Turck http://turck.com/

Software

HMI, Programming, Control and Design Software

Autodesk http://usa.autodesk.com/
Avantis (Enterprise S/W) http://iom.invensys.com/EN/Pages/Avantis.aspx
Microsoft http://www.microsoft.com/
National Instruments http://www.ni.com/
PTC (Pro/E) http://www.ptc.com/
SAP (Enterprise S/W) http://www.sap.com/index.epx
Solidworks http://www.solidworks.com/
Wonderware http://www.wonderware.com/

Vision

Vision Systems and Components

Banner http://www.bannerengineering.com/en-US/
CCS America http://www.ccsamerica.com/ (Lighting)
Cognex http://www.cognex.com/
Keyence http://www.keyence.com/
Matrox http://www.matrox.com/
Microscan http://www.microscan.com/en-us/
PPT (Datalogic) http://www.pptvision.com/

Machine Builders and Integrators

ATS Automation http://www.atsautomation.com/ (Custom machinery and integration)
Automation Consulting, LLC http://www.automationllc.com/ (Automated Systems and Consulting)
Automation nth http://www.automationnth.com/ (Controls and Integration)
Automation Tool http://www.automationtool.com/ (Test Machinery and Integration)
Bachelor Controls http://www.bachelorcontrols.com/ (System Integrators)
Concept Systems http://www.conceptsystemsinc.com/

Appendix E

Doerfer Companies	http://www.doerfer.com/
	(WrightIndustries, AdvancedAutomation, Williams-White, TDS)
	(Custom machine builders and Integrators)
DW Fritz	http://www.dwfritz.com/
	(Precision automation manufacturers)
Nalle Automation Systems (NAS)	http://nalleautomation.com/
	(Packaging Machinery)
Powerhouse Controls	http://www.powerhouse.ca/home/
Precision Automation	http://www.precisionautomationinc.com/
	(Machine Building)
Revere	http://reverecontrol.com/
TKF Conveyors	http://www.tkf.com/
	(Conveyors and Material Handling)

APPENDIX F
Thermocouples

Types

Certain combinations of alloys have become popular as industry standards. Selection of the combination is driven by cost, availability, convenience, melting point, chemical properties, stability, and output. Different types are best suited for different applications. They are usually selected based on the temperature range and sensitivity needed. Thermocouples with low sensitivities (B, R, and S types) have correspondingly lower resolutions. Other selection criteria include the inertness of the thermocouple material and whether it is magnetic or not. Standard thermocouple types are listed below with the positive electrode first, followed by the negative electrode.

K

Type K (chromel{90 percent nickel and 10 percent chromium}–alumel) (Alumel consisting of 95 percent nickel, 2 percent manganese, 2 percent aluminum, and 1 percent silicon) is the most common general purpose thermocouple with a sensitivity of approximately 41 µV/°C, chromel positive relative to alumel. It is inexpensive, and a wide variety of probes are available in its −200°C to +1350°C / −328°F to +2462°F range. Type K was specified at a time when metallurgy was less advanced than it is today, and consequently characteristics may vary considerably between samples. One of the constituent metals, nickel, is magnetic; a characteristic of thermocouples made with magnetic material is that they undergo a deviation in output when the material reaches its Curie point; this occurs for type K thermocouples at around 150°C.

E

Type E (chromel–constantan) has a high output (68 µV/°C), which makes it well suited to cryogenic use. Additionally, it is nonmagnetic.

J

Type J (iron–constantan) has a more restricted range than type K (−40°C to +750°C), but higher sensitivity of about 55 µV/°C. The Curie point of the iron (770°C)[1] causes an abrupt change in the characteristic, which determines the upper temperature limit.

N

Type N (nicrosil–nisil) (nickel-chromium-silicon/nickel-silicon) thermocouples are suitable for use at high temperatures, exceeding 1200°C, because of their stability and ability to resist high temperature oxidation. Sensitivity is about 39 µV/°C at 900°C, slightly lower than type K. Designed to be an improved type K because of increased stability at higher temperatures, it is becoming more popular, although the differences may or may not be substantial enough to warrant a change.

Platinum (Types B, R, and S)

Types B, R, and S thermocouples use platinum or a platinum–rhodium alloy for each conductor. These are among the most stable thermocouples but have lower sensitivity than other types, approximately 10 µV/°C. Type B, R, and S thermocouples are usually used only for high temperature measurements because of their high cost and low sensitivity.

B

Type B thermocouples use a platinum–rhodium alloy for each conductor. One conductor contains 30 percent rhodium, while the other conductor contains 6 percent rhodium. These thermocouples are suited for use at up to 1800°C. Type B thermocouples produce the same output at 0°C and 42°C, limiting their use below about 50°C.

R

Type R thermocouples use a platinum–rhodium alloy containing 13 percent rhodium for one conductor and pure platinum for the other conductor. Type R thermocouples are used up to 1600°C.

S

Type S thermocouples are constructed using one wire of 90 percent platinum and 10 percent rhodium (the positive or "+" wire) and a second wire of 100 percent platinum (the negative or "−" wire). Like type R, type S thermocouples are used up to 1600°C. In particular, type S is used as the standard of calibration for the melting point of gold (1064.43°C).

T

Type T (copper–constantan) thermocouples are suited for measurements in the −200 to 350°C range. Often used as a differential measurement since only copper wire touches the probes. Since both conductors are nonmagnetic, there is no Curie point and thus no abrupt change in characteristics. Type T thermocouples have a sensitivity of about 43 µV/°C.

C

Type C (tungsten 5 percent rhenium–tungsten, 26 percent rhenium) thermocouples are suited for measurements in the 0°C to 2320°C range. This thermocouple is well suited for vacuum furnaces at extremely high temperatures. It must never be used in the presence of oxygen at temperatures above 260°C.

M

Type M thermocouples use a nickel alloy for each wire. The positive wire (20 alloy) contains 18 percent molybdenum, while the negative wire (19 alloy) contains 0.8 percent cobalt. These thermocouples are used in vacuum furnaces for the same reasons as with type C. Upper temperature is limited to 1400°C. It is less commonly used than other types.

Chromel–Gold/iron

In chromel–gold/iron thermocouples, the positive wire is chromel and the negative wire is gold with a small fraction (0.03 to 0.15 atom percent) of iron. It can be used for cryogenic applications (1.2 to 300 K and even up to 600 K). Both the sensitivity and the temperature range depends on the iron concentration. The sensitivity is typically around 15 µV/K at low temperatures, and the lowest usable temperature varies between 1.2 and 4.2 K.

Appendix F

Type	Temperature Range (Centigrade)	ASTM Letter	Element Alloys	ASTM Colors	IEC Colors	Japan JIS Colors
B	+200 to +1700	B+	Pt 30% Rh			Gray
		B–	Pt 6% Rh			Red
E	0 to +800	E+	Chromel	Purple	Purple	Red
		E–	Constantan	Red	White	White
J	0 to +750	J+	Iron	White	Black	Red
		J–	Constantan	Red	White	White
K	0 to +1100	K+	Chromel	Yellow	Green	Red
		K–	Alumel	Red	White	White
N	0 to +1100	N+	Nicrosil	Orange	Pink	
		N–	Nisil	Red	White	
R	0 to +1600	R+	Pt 13% Rh		Orange	Red
		R–	Pure Pt		White	White
S	0 to +1600	S+	Pt 10% Rh		Orange	Red
		S–	Pure Pt		White	White
T	–185 to +300	T+	Copper	Blue	Brown	Red
		T–	Constantan	Red	White	White

TABLE F1 Thermocouples

Bibliography

"About the HART Protocol," HART Communication Foundation, www.hartcomm.org.

Blackburn, J. A., *Modern Instrumentation for Scientists and Engineers*, Springer-Verlag, New York, 2001.

Brown, H. T., *507 Mechanical Movements, Mechanisms and Devices*, Dover Publications, Mineola, NY, 2005.

Bruce, R. G., Dalton, W. K., Neely, J. E., and Kibbe, R. R., *Modern Materials and Manufacturing Processes*, Prentice Hall, Boston, 1998.

Chapra, S. C., and Canale, R. P., *Introduction to Computing for Engineers*, McGraw-Hill College, New York, 1986.

Craig, J. J., *Introduction to Robotics, Mechanics and Control*, Addison-Wesley, Boston, 1989.

Downs, B. T., and Grout, J. R., *A Brief Tutorial on Mistake-Proofing, Poka-Yoke and ZQC*, http://facultyweb.berry.edu/jgrout/tutorial.html, pdf document.

George, M. L., *Lean Six Sigma: Combining Six Sigma Quality with Lean Speed*, McGraw-Hill, New York, 2002.

Harry, M., and Schroeder, R., *Six Sigma: The Breakthrough Management Strategy Revolutionizing the World's Top Corporations*, Double Day, New York, 2000.

Laughton, M. A., and Warne D. F., *Programmable Controller. Electrical Engineer's Reference Book*, 16th ed., Newnes, Oxford, 2003.

Newell, M. W., and Grashina, M. N., *The Project Management Question and Answer Book*, AMACOM, New York, 2004.

Oberg, E., Jones, F. D., Horton, H. L., and Ryffel, H. H., *Machinery's Handbook*, 27th ed., Industrial Press, New York, 2005.

Pallante, R., *Application Equipment for Cold Adhesives*, http://www.nordson.com/en-us/divisions/adhesive-dispensing/Literature/PKR/PKR1644.pdf, pdf document.

Paul, R. R., *Robot Manipulators: Mathematics, Programming, and Control*, MIT Press, Cambridge, MA, 1981.

Sen, P. C., *Principles of Electric Machines and Power Electronics*, John Wiley and Sons, New York, 1989.

Smith, W. F., *Principles of Materials Science and Engineering*, McGraw-Hill College, New York, 1990.

Spiteri, C. J., *Robotics Technology*, Saunders College Pub., Philadelphia, 1990.

Stevenson, W. J., *Operations Management*, McGraw-Hill/Irwin, New York, 2007.

Thorne, M., *Computer Organization and Assembly Language Programming*, Krieger Publishing, Malabar, FL, 1991.

Vermaat, S. C., *Discovering Computers 2008*, Thomson Course Technology, Boston, 2008.

Zuch, E. L., *Data Acquisition and Conversion Handbook*, Datel Intersil, Mansfield, MA, 1979.

Index

Note: Section numbers are in brackets followed by page numbers.

3D Modeling [6.2], 240
5S [8.3.4], 279
80–20 [3.8.2, 10.1, App. E], 153, 305, 337

A

ABB [App. E], 338
AC [2.4], 25
Accumulator [5.4], 191
Accuracy [4.4.5], 182
Actuator [3.5], 115
ADC [2.1], 9
Additive Manufacturing (3D Printing) [1.1.3], 5
Air Logic [2.2.1], 12
Allen-Bradley [App. E], 338
Analog [2.1], 9
Anodizing [5.5.1], 198
ASCII [2.3.6, App. A], 24, 317
ASIBus [2.2.3], 20
Assembly [7.2.1], 258
Asynchronous [2.6.2], 36
AutoCAD [2.7.1, App. E], 37, 339
Automation [1.1], 1
Automation Direct [App. E], 338

B

Baldor [App. E], 337
Banner [App. E], 338
Barber Colman [App. E], 336
Bar Code [3.3.3], 90
Barriers [2.8.8], 56
Bearing [3.7.4], 141
Belt [4.1.1], 166
Binary [2.3.1], 22
BIST [2.8.7], 55
Bit [2.1, 2.3.1], 10, 22
Bit-Banging [9.4.2], 295
Bolt [3.8.1], 152
BOOL [2.3.1, 6.1.3], 22, 227
Bosch [App. E], 337
Bracket [3.8.2], 153
Butt Splice [3.4.7], 111
Byte [2.3.5], 24

C

Cable [3.4.7], 110
Cam [3.7.1], 135
Camco [4.2.1, App. E], 172, 337
CAN [2.2.3], 19
CanOPEN [2.2.3, 3.5.3], 20, 120
Capacitive [3.3.1], 75
Cartesian [4.4.3], 180
CE (Communite' Europe'ene) [2.8], 45
Chemical [5.1, 7.1.4], 186, 254
CIP [2.2.3], 18
Circuit breaker [3.4.1], 94

348 Index

Clean room [1.1.4], 6
Color [3.3.1, 3.3.3], 71, 80
Communications [2.2.3], 15
Compliance (robot) [4.4.5], 182
Computer [3.1.1], 61
Contactor [3.4.4], 104
Continuous [2.6.1], 36
Controls [7.1.2], 252
Converting [5.4], 192
Conveyor [4.1], 165
Counter [3.4.5, 6.1.3], 107, 228
CPU [3.1.4], 64
Creform [3.8.3, App. E], 153, 337
Current [2.4.2], 25

D

DAC [2.1], 9
DC [2.4], 25
DCS [3.1.2], 62
Debounce [3.3.1], 71
Debug [6.1.2, 7.1.2], 224, 252
Decimal [2.3.2], 22
Decoder [2.2.4], 21
Designer [7.1.5], 256
DeviceNet [2.2.3], 19
DHCP [2.2.3], 18
Digital [2.1], 9
Dimensioning [2.7.1], 38
DIN [3.1.4, 3.4.2], 64, 99
DIN rail [3.4.2], 99
DINT [2.3.5, 6.2.4], 24, 227
Disconnect [2.8.4, 3.4.1], 51, 94
Discrete [2.2.1], 11
Distance measurement [3.3.2], 81
Distributed I/O [2.2.3, 3.1.2], 15, 62
Distribution block [3.4.2], 98
DLR (Device Level Ring) [2.2.3], 19
Dowel [3.8.1], 149
Drain [3.4.7], 111
Dry Cycle [9.6.3], 302

E

Electrical engineer [7.1.2], 252
Electrician [7.2.2], 261
Emergency stop [2.8.2], 48

Emerson [10.1, App. E], 306, 337
Enclosures [3.8.4], 154
Encoder [3.3.3], 84
Engineering [7.1], 251
EtherCAT [3.5.3], 120
Ethernet [2.2.3], 17
Ethernet/IP [2.2.3], 18
Ethernet Powerlink [3.5.3], 120
Extraction [5.5.1], 192
Extrusion [3.8.2, 5.5.1, 5.5.2], 153, 194, 199

F

Factory [1.1.3], 3
Fastener [3.8.1], 151
Factory Acceptance Testing (FAT) [9.7.1], 302
FEM (Finite Element Method) [9.4.1], 293
Ferrule [3.4.6], 112
Festo [App. E], 338
Fieldbus [2.2.3], 20
Fitting [2.5.1, 3.4.7], 33, 111
Fixture key [3.8.1], 151
Floating point [2.3.4, 6.1.3], 24, 227
Fluid power [2.5], 32
FMEA [8.2.3], 268
Food and beverage [5.2], 187
Frame grabber [3.3.3], 88
Framing [3.8], 148
Fuseblock [3.4.2], 99
Fusing [3.4.1], 97

G

Galvanizing [5.5.1], 198
Gantry robot [4.4.3], 180
Gauging [3.3.2, 8.1.6], 77
GE (General Electric) [App. E], 338
GE Fanuc [App. E], 338
Gearing [3.7.3], 136
Geometric tolerancing [2.7.1], 38
Graceport [9.6.3], 301
Grind spacer [3.8.1], 151
Guarding [2.8.3, 3.8.2, 9.4.1], 51, 153, 294
GUI [3.2], 66

Index

H

HART Protocol [2.2.3], 20
Heat shrink [3.4.7], 111
Heijunka box [8.3.1], 275
Hexadecimal [2.3.3], 22
High speed counter [2.2.4, 3.3.3], 21, 85
HMI [3.2], 66
Hoffman [3.8.4, App. E], 155, 336
Homing [2.2.4, 3.6.4], 21, 130
Horsepower [2.4.3], 27
Hydraulics [2.5.2], 34
Hydroforming [5.5.1], 197

I

I/O [2.2], 11
Idec [App. E], 336
Imaging [2.7.3, 3.3.3], 42, 87
Indexer [4.2], 172
Inductive [3.3.1], 74
Industrial engineer [7.1.3], 253
Infrared thermocouple [3.3.2], 84
Instrumentation [3.3.2, 7.2.2], 79, 262
Integration [5.0], 186
Integrator [8.1.6], 265
Intrinsically Safe [2.2.1, 2.8.8], 11, 56
Ishikawa (fishbone) diagram [8.3.4], 280, 281
ISO [8.2.7], 271
Isolation transformer [3.4.3], 101
Item [3.8.2, App. E], 153, 337

J

JIT (Just In Time) [8.3], 274
Jumper [3.4.2], 99

K

Kaizen [8.3.2], 276
Kanban [8.3.1], 275
Keyboard wedge [3.3.3], 92
Keyence [App. E], 339
KW [2.4.3], 27

L

Ladder logic [3.1.3, 6.2.4], 64, 221, 228
LAN (Local Area Network) [2.2.3], 17
Lean manufacturing [1.1.3, 8.3], 4, 273
Leveling feet [3.8.1], 148
Linear bearing [3.7.4], 141
Linearity [2.1], 10
Linear motor [3.6.3], 129
Load cell [3.3.2], 78
LVDT [3.3.2], 80

M

Machine builder [8.1.5], 264
Machine pad [3.8.1], 149
Machinery directive [2.8], 45
Machining [5.5.1, 7.2.1], 197, 251
Magnetostrictive [3.3.2], 82
MCR (Master Control Relay) [2.8.2], 51
Measurement [3.3.2], 77
Mechanical engineer [7.1.1], 252
Mechanism [3.7], 134
Meter In/Out (Flow Controls) [2.5.1], 33
Microsoft [App. E], 339
Millwright [7.2.1], 259
Misumi [3.8.3, App. E], 154, 337
Mitsubishi [App. E], 338
Modbus [2.2.3], 20
Modicon [App. E], 338
Monitor [3.2], 65
Motion control [3.5.3], 119
Motor [3.6], 121
Motor Starter [3.4.4], 104
MSDS (Material Safety Data Sheet) [1.1.4], 8
MTS Temposonics [3.3.2], 82
Muda [8.3], 273
Mura [8.3], 274
Muri [8.3], 275
Multiconductor [3.4.7], 110

350 Index

N

Namur [2.8.8], 57
National Instruments [App. E], 338
Network [2.2.3], 15
Network security [6.7], 248
Node [2.2.3], 15

O

Octal [2.3.3], 22
OEE (Overall Equipment Effectiveness) [2.9, 6.6, 8.3.4], 57, 243, 279
Off delay [3.4.5], 104
Office software [6.4], 241
OIT (Operator Interface Terminal) [3.2], 65
Omron [App. E], 338
On delay [3.4.5], 104
One Shot [3.4.5], 104
OSHA [2.8], 46
Overload [3.4.4], 104
Overtravel [3.6.4], 130

P

P&ID [2.7.2], 40
Packaging [5.3], 188
Panelbuilding [7.2.2, 9.5.3], 260, 298
Parallel [2.2.3], 17
Pepperl+Fuchs [2.8.8, App. E], 57, 336
Phase [2.4.4], 27
Phoenix Contact [App. E], 336
Photoeye [3.3.1], 71
Pick and place [4.2.4, 6.1.3], 174, 237
PID [2.2.2, 3.1.4], 13, 64
Piping [3.8.3], 153
PLC [3.1.3], 62
Pneumatics [2.5.1], 33
Pokayoke [8.3.3], 277
Power [2.4.3, 3.4], 27, 93
Power factor [3.6.1], 123
Power supply [3.4.3], 102
PPAP (Production Part Approval Process) [8.2.3], 263

Pro E [App. E], 339
Process control [5.0], 185
Profibus [2.2.3, 9.4.2], 20, 295
ProfiNET IRT [3.5.3], 120
Programming [6.1.2], 221
Project manager [7.1.5], 257
Proportional valve [5.1], 186
Proximity switches [3.3.1], 74
Pulley [3.7.4], 142
Pulse [3.4.5], 104

Q

QMS (Quality Management System) [8.2.7], 271
Quadrature [3.3.3], 85
Quick disconnect [3.3.1], 70

R

Ratchet and pawl [3.7.2], 136
Reactor [5.1], 187
REAL [2.3.4, 6.2.4], 24, 227
Relay [3.4.4], 102
Repeatability [4.4.5], 182
Resolution [2.1], 9
Resolver [3.3.3], 86
RFID [3.3.3], 91
Rigger [7.2.1], 259
Rittal [App. E], 336
Rivet [3.8.1], 152
Robot [4.4], 178
RoHS (Restriction of Hazardous Substances) [1.1.4, 2.8], 8, 44
Roll [5.5], 194
Rollforming [5.5.1], 194
Rotor [3.6.1], 121
RS232 [2.2.3], 16
RS422/485 [2.2.3], 17
RTD [3.3.2], 84

S

Safety [2.8], 44
SAT (Site Acceptance Test) [9.7.2], 302
SCADA [6.5], 242
Scaling [2.1.1], 10
Scanning [6.1.3], 228
SCARA [4.4.2], 179

Index

T

TCP/IP [2.2.3], 18
Teach pendant [4.4.5], 182
Temperature [3.3.2], 82
Temperature controller [3.1.4], 64
Template [9.4.3], 296
Terminal block [3.4.2], 98
Thermister [3.3.2], 84
Thermocouple [3.3.2], 82
Three phase [2.4.4], 27
Timer [3.4.5], 104
Tolerance [2.7.1], 38
Topology [2.2.3], 15
Touchscreen [3.2.3], 67
TPM (Total Productive Maintenance) [8.2.4], 269
TQM (Total Quality Management) [8.2.7], 272
Transformer [3.4.3], 100
Transducer [3.3.2], 77
Turck [App. E], 339
Twisted pair [2.2.3, 3.4.7], 16, 110

U

UL (Underwriter's Laboratories) [2.8, 3.4.1], 46, 96
Ultrasonic [3.3.2], 81
USB [2.2.3], 19

V

Valve [3.5.1, 5.1], 117, 186
Variable [6.1.1], 220
Vibratory bowl [4.3.1], 175
Vision system [3.3.3], 86
Voltage [2.4.2], 25
VSM (Value Stream Mapping) [8.3.4], 279

W

Walking beam [4.2.3], 173
Watt [2.4.3], 27
Web [5.4], 190
Weight [3.3.2], 78
Welding [5.6.2, 7.2.1], 210, 259
Wire [3.4, 3.4.7], 94, 110
Wire EDM [5.5.1], 197
Wireless [2.2.3], 21
WLAN (Wireless LAN) [2.2.3], 21
Word [2.3.5], 24

Screw [3.8.1], 151
SDS (Safety Data Sheet) [1.1.4], 8
Sequencer [6.1.3, 9.4.3], 228, 293
Sensor [3.3], 69
SERCOS [3.5.3], 120
Serial [2.2.3], 16
Servo [3.5.3, 3.6.4], 119, 129
Servomechanism [3.7.5], 143
Shield [3.4.7], 111
Shim [3.8.1], 149
Siemens [8.1.4, 9.4.2, App. E], 264, 295, 338
Simple machine [3.7], 134
Sinking [3.3.1], 70
Sintering [5.5.1], 197
Six Sigma [1.1.3, 8.2.7], 4, 272
Slip [3.6.1], 122
Slitter [5.4], 192
SMC [App. E], 338
Smelting [5.5.1], 192
Software [6.0], 219
Soldering [3.4.6], 112
Solenoid [3.5.1], 117
Solid modeling [2.7.1], 37
Solidworks [App. E], 339
Sourcing [3.3.1], 70
SPC (Statistical Process Control) [8.3.4], 279
Splice [3.4.7], 111
Spline [3.7.3], 136
Statement of work [9.2.3], 291
Stator [3.6], 121
Stelron [App. E], 337
Stepfeeder [4.3.2], 176
Stepper motor [3.6.4], 131
Strain gauge [3.3.2], 77
Stress relieving [3.8.1], 149
STRING [2.3.6, 6.2.4], 24, 227
SWI (Standardized Work Instructions) [8.3.4], 278
Synchronous [2.6.2], 36
Systems engineer [7.1.5], 255
Systems integration [8.1.6], 265

X

X-Ray [3.3], 89

Y

Yaskawa [App. E], 337

Z

Z pulse [3.3], 85